SpringerBriefs in Applied Sciences and Technology

SpringerBriefs present concise summaries of cutting-edge research and practical applications across a wide spectrum of fields. Featuring compact volumes of 50–125 pages, the series covers a range of content from professional to academic.

Typical publications can be:

- A timely report of state-of-the art methods
- An introduction to or a manual for the application of mathematical or computer techniques
- A bridge between new research results, as published in journal articles
- A snapshot of a hot or emerging topic
- An in-depth case study
- A presentation of core concepts that students must understand in order to make independent contributions

SpringerBriefs are characterized by fast, global electronic dissemination, standard publishing contracts, standardized manuscript preparation and formatting guidelines, and expedited production schedules.

On the one hand, **SpringerBriefs in Applied Sciences and Technology** are devoted to the publication of fundamentals and applications within the different classical engineering disciplines as well as in interdisciplinary fields that recently emerged between these areas. On the other hand, as the boundary separating fundamental research and applied technology is more and more dissolving, this series is particularly open to trans-disciplinary topics between fundamental science and engineering.

Indexed by EI-Compendex, SCOPUS and Springerlink.

More information about this series at http://www.springer.com/series/8884

Aydin Azizi · Poorya Ghafoorpoor Yazdi

Computer-Based Analysis of the Stochastic Stability of Mechanical Structures Driven by White and Colored Noise

Springer

Aydin Azizi
Department of Engineering
German University of Technology
in Oman
Muscat, Oman

Poorya Ghafoorpoor Yazdi
Department of Engineering
German University of Technology
in Oman
Muscat, Oman

ISSN 2191-530X   ISSN 2191-5318   (electronic)
SpringerBriefs in Applied Sciences and Technology
ISBN 978-981-13-6217-0   ISBN 978-981-13-6218-7   (eBook)
https://doi.org/10.1007/978-981-13-6218-7

Library of Congress Control Number: 2019930372

This Springer imprint is published by the registered company Springer Nature Singapore Pte Ltd.
The registered company address is: 152 Beach Road, #21-01/04 Gateway East, Singapore 189721, Singapore

Acknowledgements

This work was supported by The Research Council of Oman with the grant number ORG/CBS/14/008.

Contents

1 Introduction to Fuel Consumption Optimization Techniques 1
   1.1 Overview 1
      1.1.1 Factors Effecting Fuel Efficiency 2
   1.2 Formulation of the Problem 5
      1.2.1 Methods of Improving the Fuel Efficiency 5
      1.2.2 Methods of Improving the Vehicle Suspension 6
   1.3 Scope and Contribution 7
   1.4 Methodology 8
   1.5 Organization of the Book 8
   References 9

2 Introduction to Noise and its Applications 13
   2.1 Overview 13
   2.2 Background and History of Noise 13
   2.3 Noise in Electronic View 14
   2.4 Noise in Communication View 14
   2.5 Different Types of Noise 14
      2.5.1 Typical Types of Noise 15
      2.5.2 Coupled Type of Noise 17
      2.5.3 Colored Noise in Signal Processing View 20
   References 22

3 White Noise: Applications and Mathematical Modeling 25
   3.1 Overview 25
   3.2 Practical and Real-Life Applications of White Noise 25
      3.2.1 White Noise in Electronics Engineering 25
      3.2.2 White Noise in Building Acoustics 26
      3.2.3 White Noise for Tinnitus Treatment 26
      3.2.4 White Noise to Improve the Work Environment 26

3.2.5   White Noise in Earthquake Simulation   26
3.2.6   Mathematical Modeling of Gaussian White Noise
        as Pavement Condition   32
References   36

4   **Mechanical Structures: Mathematical Modeling**   37
    4.1   Overview.   37
          4.1.1   Degrees of Freedom.   37
          4.1.2   Periodical System Response   38
          4.1.3   Harmonic Motion   38
          4.1.4   Frequency   39
          4.1.5   Amplitude   39
          4.1.6   The Mean Square Amplitude   39
          4.1.7   Free and Forced Vibrations   40
          4.1.8   Phasor.   40
    4.2   Single Degree of Freedom Mechanical Systems   41
    4.3   SODF Free Motion.   42
    4.4   Damped Motion of Single Degree of Freedom Systems   44
    4.5   Forced Response of SDOF Systems.   46
    4.6   Car Suspension Types as Mechanical Structures
          and Performance.   49
          4.6.1   Passive Suspension   49
          4.6.2   Semi-active Suspension   50
          4.6.3   Active Suspension   51
          4.6.4   Mathematical Modeling of Passive Suspension System . . .   53
          4.6.5   Mathematical Modeling Active Suspension System   55
    References   57

5   **Noise Control Techniques**   61
    5.1   Overview.   61
    5.2   Adaptive Noise Filter   61
    5.3   Adoptive Noise Cancelation   65
    5.4   PID Controller   68
          5.4.1   P Controller   68
          5.4.2   I Controller   69
          5.4.3   D Controller   69
          5.4.4   PI Controller   69
          5.4.5   PD Control   70
          5.4.6   Mathematical Modeling of the PID Controller   70
    5.5   Sliding Mode Control   71
          5.5.1   Mathematical Model of the Sliding Mode Controller   71
    References   71

6 Modeling and Control of the Effect of the Noise on the Mechanical Structures 75
   6.1 Overview 75
   6.2 Active Suspension System 75
   6.3 Design of PID Controller 77
      6.3.1 Results 79
   6.4 Design of Sliding Mode Controller 83
      6.4.1 Stability in the Sense of Lyapunov 84
      6.4.2 Asymptotic Model 84
      6.4.3 Lyapunov's Direct/Second Method 85
      6.4.4 Conclusion 91
   References 92

List of Figures

Fig. 2.1   Different types of noise categories 15

Fig. 3.1   Simulink model of the signal of White noise pavement
roughness 34

Fig. 3.2   Simulink model of Class C pavement roughness signal 35

Fig. 4.1   Harmonic motion's phasor diagram 40

Fig. 4.2   Undamped and damped single degree of freedom
systems 42

Fig. 4.3   Different value of damping and their response 45

Fig. 4.4   Magnitude of forced response of a single degree of freedom
system 48

Fig. 4.5   Phases of forced response of a single degree of freedom
system 48

Fig. 4.6   Passive suspension system 50

Fig. 4.7   Semi-active suspension system 51

Fig. 4.8   ¼ model of active suspension system 52

Fig. 4.9   Vehicle ¼ passive suspension system mathematical
model 53

Fig. 4.10 ¼ model of active suspension system 56

Fig. 5.1   General block diagram of adaptive filter 62

Fig. 5.2   FIR filter structure 63

Fig. 5.3   Structure of the infinite impulse response filter 64

Fig. 5.4   Concept of ANC 67

Fig. 5.5   PID controller 68

Fig. 6.1   Proposed quarter car model equipped with Active
suspension system 76

Fig. 6.2   Active suspension system equipped with PID controller 78

Fig. 6.3   Block diagram of the proposed model equipped
with PID controller 78

Fig. 6.4   MATLAB–Simulink model of the noise cancelation system
for the active suspension system 79

Fig. 6.5   Uncontrolled total system response based on the random
              input.   80
Fig. 6.6   Controlled total system response based on the random
              input.   81
Fig. 6.7   Controlled and uncontrolled total system behaviors   81
Fig. 6.8   Pole and zero of the designed PID controller   82
Fig. 6.9   Root locus diagram.   82
Fig. 6.10   Controlled and uncontrolled system behaviors
              for $\eta = 1$ and $\lambda = 1$ 89
Fig. 6.11   Controlled system behavior for $\lambda = 0.1$, $\lambda = 1$, $\lambda = 10$ 90
Fig. 6.12   Controlled system behavior for $\lambda = 10$, $\lambda = 100$, $\lambda = 1000$. . . .   90
Fig. 6.13   Controlled system behavior for $\eta = 0.01$, $\eta = 0.1$, $\eta = 1$.   91
Fig. 6.14   Controlled system behavior for $\eta = 10$, $\eta = 100$, $\eta = 1000$. . . .   91
Fig. 6.15   Controlled system behavior for $\eta = 1000$, $\eta = 10{,}000$,
              $\eta = 100{,}000$.   92

List of Tables

Table 2.1 Colored noise and frequencies 20
Table 3.1 Road roughness standard 35
Table 4.1 Three types of suspension systems' technical
           comparison. 52
Table 4.2 Vehicle ¼ passive suspension model simulation
           input parameters. 54
Table 5.1 Response of PID controller 70
Table 6.1 Basic theorem of Lyapunov 87

Abstract

The goal of this book is investigating and simulating the effect of white and colored noise on mechanical structures and designing effective controllers to reduce or eliminate these effects. In this book, quarter car model has been introduced as the mechanical structure, and the stochastic effect of the Gaussian white noise as the pavement condition has been modeled mathematically. To eliminate the undesired stochastic behavior of the vehicle imposed to the noise, an active suspension system has been designed, and based on the control engineering view, the mathematical model of the effect of the noise on the designed system has been driven and simulated. As the active force generators of the designed active suspension system, PID and sliding mode controllers have been introduced and designed. MATLAB software has been utilized to design the controllers as well as to model the effect of the noise on the mechanical structure and investigate the stability of the system. The results show that the designed controllers have effective performances to eliminate the effect of the noise as road condition which has a significant effect on reducing the fuel consumption and contributes to environmental sustainability.

Keywords Mechanical structure · Stochastic behavior ·
Active suspension system · Sliding mode controller · PID controller ·
White noise · Colored noise · Noise cancelation · Mathematical modeling ·
Control engineering

Chapter 1
Introduction to Fuel Consumption Optimization Techniques

1.1 Overview

Efforts to optimize fuel consumption have driven and inspired various industries, including the automobile industry, to create a wealth of new inventions and technologies. Since the issue of global warming was brought into the spotlight, the mechanics of the automobile industry have evolved rapidly, due to the greenhouse gas emissions produced by internal combustion engines. The advancement of technology within the power industry has helped in reducing fuel consumption, as well as in the reduction of greenhouse gas emissions [1].

Greenhouse gases released by vehicles include carbon dioxide, carbon monoxide, nitrogen dioxide, ozone and methane, among others. The repercussions of burning fossil fuels amount to more than just a foul smell; the aftermath of these greenhouse gases impacts the health of humans, animals, and plants alike, thus disturbing the environment and its inhabitants [2]. Within the environment, greenhouse gases disrupt the biogeochemical cycles that exist in nature resulting in problems, such as temperature rise, erosion, and droughts. One problem is the melting ice caps; even a minute temperature rise can result in rising water levels, which also increases the number of natural disasters occurring in various areas. A rise in water levels also promises depletion in land mass, due to water levels overflowing, and swallowing coastal areas. Furthermore, car emissions, along with many by-products of plants, cause the radiation from the sun to be trapped inside the Earth's atmosphere, resulting in the overall raising of the temperature. The aforementioned consequences develop into bigger problems, such as those which can already be observed in the La Nina and El Nino phenomena, and events in the Atlantic, as well as increased cyclone activity in the Indian Ocean [3].

© The Author(s), under exclusive license to Springer Nature Singapore Pte Ltd. 2019
A. Azizi and P. G. Yazdi, *Computer-Based Analysis of the Stochastic Stability of Mechanical Structures Driven by White and Colored Noise*, SpringerBriefs in Applied Sciences and Technology, https://doi.org/10.1007/978-981-13-6218-7_1

Regarding the impact on humans, these greenhouse gases play an important role in the climate change that is affecting the entire globe (global warming) and may threaten the welfare of humans physically and economically. For example, when ozone levels increase in lower elevations, it can have a direct impact on human health, including harming the respiratory system. The economy can also be affected, because of individuals who suffer from health problems caused by these issues. Furthermore, as the automobile industry grows, the use of fossil fuels will grow exponentially, bringing closer the possibility of a future with no fossil fuels, which will result in the economic downfall of bigger entities such as countries. Other greenhouse gases, such as fluorinated gases, do not interfere directly with human health, but do hurt the environment greatly, and on different levels [4].

Due to the rate at which fossil fuels are used annually (11 billion tons of oil and four billion tons of crude oil, per annum), oil deposits on Earth are predicted to run out by 2052. Furthermore, compensating the energy deficit of the oil deposits through using natural gas will only extend the lifetime of fossil fuel energy by an additional eight years. After this, the only remaining form of fossil fuel energy left would be coal. To fill in the energy gap of both the oil and natural gas deposits, coal would be so extensive we run out of fossil fuels by the year of 2088 [5]. Fossil fuels have, thus far, been the main source of energy, but with the passing of time, different sources of energy alternatives have been developed to prevent the consequences that arise from using just fossil fuels. For instance, instead of having only internal combustion engine vehicles, there are a variety of eco-friendly vehicles available, which are being used instead. Some of these eco-friendly energy sources include electronically powered vehicles, hybrid vehicles (using more than one source of energy), compressed air, etc. [6]. There are many of problems that come with the use of fossil fuels, out of which the issues with the greatest impact are its scarcity and the cost it imposes on the planet. Fossil fuels are the only plausible option for many vital functions and processes; the most important of these is transportation. Therefore, using this source of energy wisely and as efficiently as possible is a must [7].

1.1.1   Factors Effecting Fuel Efficiency

Fuel efficiency can be described as how well the chemical energy of the fuel in question is converted into kinetic energy, in terms of powertrains. Fuel efficiency varies according to several factors, including the application, the size of the vehicle, the vehicles design, power, engine parameters, and many others. Factors that affect fuel consumption are design-related, environment-related, and motorist driving strategy-related factors. Together, and individually, these factors determine how much energy needs to move the vehicle [8, 9].

1.1.1.1 Vehicle Design Factors

To design a vehicle with the lowest fuel consumption rate (irrespective of the energy source) is one of the main goals of modern car manufacturers. Car manufacturers have expended great effort on developing, and successfully applying, different approaches to a vehicle's design (especially the vehicle's size) in order to reduce its fuel consumption, as well as the harmful gases emitted. There are several parameters falling under this design category, including the aerodynamic drag, engine parameters, rolling resistance, load, and fuel type. The size of the engine refers to how much fuel can be pumped into the engine, in order to be burned and converted into energy. Meaning, the larger the capacity of the engine, the more power the car has (although that oversimplifies the concept). When a car has an engine with a capacity of two liters for instance, this means that the total amount of fuel that can fit into the cylinders is two liters, regardless of the number of cylinders. That power is measured in horsepower, or brake horsepower [10, 11].

In terms of fuel efficiency, a bigger engine does not always mean worse fuel economy. For instance, a car with a larger engine running at a high speed for a long time will use less fuel than a car with a smaller engine running at the same speed for the same length of time. Furthermore, considering recent technology, it is now made possible for a large engine, of for example six cylinders, to use only three cylinders when that is all that is needed, therefore greatly boosting fuel efficiency [12].

The aerodynamic aspect of design is concerned with improving the cars' exterior to better tackle drag, and to cause lower resistance at higher speeds. At a speed of 50 km/h, the power needed from the engine to overcome air friction is not more than 40% of the engine's power, yet when it comes to speeds greater than 80–90 km/h, the power needed increases to 60%, and more of the engine's power. This is due to proportional increasement of the drag to the square of the speed, and the power needed is proportional to the cube of the speed [13].

Rolling resistance, or rolling friction, is the resistance against the motion of a tire when rolling on a certain surface. Three forms of rolling resistance are in question, namely: permanent deformation, hysteresis losses, and slippage between the surface and the tire. Interestingly, a worn-out tire gives lower rolling resistance than a new one, due to the depth of the tread on the tire and its friction-inducing properties which in turn, leads to lower fuel consumption. Hysteresis losses are the main reason for rolling friction. Because tires are made of a deformable material, the tire is subject to repeated cycles of deformation and recovery, there is energy dissipated as heat. The energy is lost basically when the energy of healing/recovery is less than that of the deformation. Because of rubber's properties, it does not recover or heal over a short period of time, it needs a longer time. Some manufacturers include silicone in the tread of the tires to cut down on the time needed for the tire to recover, and hopefully cut down on the lost energy rates [14, 15].

Fuel type also influences fuel consumption. A diesel engine delivers better fuel economy for several reasons, the main reason being that diesel contains higher energy content than gasoline. However, that does not mean it affects performance in terms of speed. It instead delivers more power and torque, which is why most trucks have diesel engines. When comparing a diesel engine to a gasoline engine, the difference in fuel economy shows the diesel engines to be 25–30% more fuel economic, and in some cases, up to 40% more economic. The obvious cut back to using a diesel engine is speed in terms of performance, on the other hand, when compared to a gasoline engine of the same size; more economic [16]. A diesel-powered engine is a very sophisticated one and, in a sense, more delicate than a gasoline-powered engine. For example, if water somehow gets into the fuel tank (whether it is the suppliers' fault or a combination of heavy rain and bad luck), it could lead to many problems, including the most obvious one, severe damage to the engine [17, 18].

The load on the vehicle is another factor that affects fuel economy of car. When a vehicle is overloaded, the power it requires increases. This is because the engine will need to produce more power to move the vehicle at any given speed, in addition to the increased load on the tires, which in return, increases rolling resistance (which also increases fuel consumption) [19].

1.1.1.2 Environmental Factors

The weather and climate influence the fuel economy of cars although not drastically. In the winter for example, the air is heavier and denser, and so the air drag coefficient is larger. The tires experience a decrease in pressure which decreases fuel economy. Lubricants, wherever they may be, become cold and harder and interfere with fuel economy as well due to the increase in friction. In the summer, however, the air is lighter and less dense making it easier to navigate in contrast [20].

The terrain in which a car travels can influence its fuel consumption in a positive or negative manner. Rough terrain, ascents, slippery surfaces, sandy/muddy surfaces, and other types of terrain affect fuel consumption negatively. The opposite of these terrains can either not affect the fuel consumption or even affect it positively. For instance, climbing up a hill will require a great deal more power which means burning more fuel while descending a hill will require little or no fuel to be used [21].

Driving within a city uses much more petrol per km when compared to driving on a highway or in the countryside. The reason for that is the limitations and conditions that exist in the city but not outside it or on the highway such as traffic, speed bumps, more turns, and pedestrians. When driving in the city, the driver must use the brakes a greater number of times decreasing fuel economy. Another active factor is the fact that in the city, the driver is forced to drive at lower speeds. Speeds lower than the optimum speed also affects fuel consumption and significantly more when combined with the reasons mentioned. When driving on the highway or the countryside, the limitations faced in the city are either not faced, or significantly lower. The vehicle is much more likely to be driven at optimum speeds leading to a better fuel consumption rate as well [22–24].

1.2 Formulation of the Problem

One of the important factors which has a great influence on fuel consumption of transportation vehicles is pavement condition [25]. It is a fact that, due to the interaction between tire and the pavement, the tires of a vehicle partially deform; this deformation results in the stored potential energy of the tires being converted to heat, which is partly absorbed by the rest of tire, with the remainder being dissipated into the atmosphere [26], so it is important to know that higher pavement texture results in more fuel consumption [27]. In past decade due the problem of global warming and drawbacks of high consumption rate, modeling and simulating the effect of different types of pavement conditions on fuel consumption rate and its effect on environment has been an interested topic for many researchers [28–32]. It is an interesting point to know that the results have continuously shown that pavement smoothness has the highest impact on the rate of fuel consumption; the smoother the road, the less fuel consumption [33–36].

When a vehicle travels in a constant speed, the pavement roughness can be considered as a stochastic process subjected to Gauss distribution [37]. In this case, since the speed power spectral density of the vehicle considered as a constant value, it matches with the statistical characteristic of the white noise, so pavement roughness model can be considered as the white noise input signal [38].

There is an energy loss during the vehicle transportation on the road. The energy is absorbed and converted to thermal energy form by suspension system and tires, so it is obvious that the energy loss will be reduced if the suspension system can eliminate the effect of the pavement condition and make vehicle bounces less [39]. In other word, it means that designing an effective suspension system to compensate the effects of pavement conditions on vehicle is one of the solutions to reduce fuel consumption.

1.2.1 Methods of Improving the Fuel Efficiency

One of the ways to decrease the load/mass of the car is choosing the high-tech materials, as to increase fuel efficiency without jeopardizing the safety of the passenger. Some automakers are trying to use plastic fuel tanks and carbon fiber instead of steel. Alas, in the case of carbon fiber which is a lightweight material that is also reliable, it is not likely to be used to due to its hefty price in the market. A record by the EPA (environmental protection agency) shows that for every 45 kg of mass reduced, fuel efficiency can increase by one to two percent (1–2%) [40, 41].

Modifying the engine appropriately (ex: adding/improving a turbo charger) can improve efficiency of fuel consumption by up to 4 percent (4%) and fixing a serious problem such as a broken oxygen sensor can improve efficiency by a staggering 40 percent (40%). Keeping the pressure of the tires in check (at the adequate pressure accordingly) improves fuel consumption by up to 3.3%. Old cars that use a carbureted engine are still in use, and by keeping their air filters unclogged could help with fuel economy and acceleration. In more modern cars, however, it usually aids with acceleration only [42–44].

One crucial step in optimizing fuel efficiency involves developing analytical models to predict the vehicle's fuel consumption and to achieve the desired results. There have been models that compute fuel consumption estimates in cars with respect to their fuel consumption, characteristics, and the surrounding environment. However, these models only represent approximations to these estimates. By adding more variables to the models, the outcome will become more accurate, but this will result in a less efficient model. This is because mathematical models represent approximations to the real result while having errors in each of the parameters used. Therefore, the more parameters one uses, the more errors are involved, thus making the model less efficient [45].

There have been many attempts to enable mathematical models to predict fuel consumption with different variables in each. For the first time to predict vehicle fuel consumption and emissions metrics, mathematical models were utilized by Ahn et al. [46]. The proposed model is a function of speed and acceleration with constant parameters and can predict fuel consumption or CO emission rates for an assumed vehicle. According to Ross et al. [47] there are two factors that affect vehicle fuel consumption: the efficiency of the powertrain and the power required in working the vehicle. He evaluated fuel consumption by finding the product of optimal specific fuel consumption into the sum of the powers of the rolling resistance, air resistance, and inertial acceleration resistance and then dividing it by the product of the efficiency of transmission with the average speed of the vehicle and the fuel density. In another research on analytic modeling of vehicle fuel consumption, it has been shown that the fuel consumption can be found by using the calorific value of fuel per 100 km [45]. It was done by finding the sum of energies of the forces required to overcome resistance and the kinetic energy required for episodic accelerations and dividing it by the calorific value of the fuel used. In 2008, Smit et al. [48] conducted a research which examined how, and to what extent, models used to predict emissions and fuel consumption from road traffic, including the effects of congestion. In 2011, Rakha et al. [49] developed a fuel consumption model which can be easily integrated within a traffic simulation framework. In 2015, Tang et al. [50] proposed a model to investigate the impacts of the driver's bounded rationality and the effect of signal lights on fuel consumption. In 2017, a fuel consumption model for heavy-duty tracks has been proposed by Wang et al. [51].

1.2.2   Methods of Improving the Vehicle Suspension

From point of view of ride safety, the most important element of the vehicle which has a direct impact on passengers comfort is suspension system [52]. Nowadays, with advancements in the car manufacturing industry, companies attempt to provide a smooth ride for passengers through the development and manufacture of more advanced vehicle suspension systems. These systems are able to minimize the effects of uneven pavement conditions of roads on the passengers [53]. In 2016, Kognati et al. [54] proposed a unified approach to model complex multibody mechanical systems and design controls for them, Pappalardo et al. [55] in 2017 proposed a novel methodology to address the problems of suppressing structural vibrations and attenuating contact forces in nonlinear mechanical systems, in 2018 they developed an adjoint method which can be effectively used for solving the optimal control problem associated with a large class of nonlinear mechanical systems [56]. Nowadays to optimize quality of traveling by cars, various types of controllers such as adaptive control [57], linear quadratic Gaussian (LQG) control [58], H-infinity [59], proportional (P) controller [60], proportional integral (PI) controller [61], and proportional–integral–derivative (PID) controller [62] have been utilized, in order to control the car suspension system and to eliminate the vibration coming from the pavement. In 2014, Li et al. [63] proposed an output feedback H∞ control for a class of active quarter car suspension systems with control delay. In 2015, AENS et al. [64] performed a comparison between passive and active suspensions systems. In 2016, Buscarino et al. [65] investigated the role of passive and active vibrations for the control of nonlinear large-scale electromechanical systems. Zhao et al. [66] in 2016 utilized adaptive neural network control for active suspension system with actuator saturation. Taskin et al. [67] in 2017 investigated the effect of utilizing fuzzy logic controller on an active suspension system based on a quarter car test rig; and in 2018 a control scheme, utilizing Hybrid ANFIS PID, was proposed by Singh et al. [68] to improve the passenger ride comfort and safety in an active quarter car model. Fauzi et al. [69] in 2018 developed a state feedback controller to reduce body deflection caused by road disturbance to achieve the ride comfort of driver and passengers.

1.3 Scope and Contribution

The goal of this research is investigating and simulating the effect of white and colored noises on mechanical structures and designing effective controllers to reduce or eliminate these effects. In this book, quarter car model has been introduced as mechanical structure, and the stochastic effect of the Gaussian white noise as the pavement condition has been modeled mathematically. To eliminate the undesired stochastic behavior of the vehicle imposed to the noise, an active suspension system has been designed and based on the control engineering view the mathematical model of the effect of the designed system on the quarter car model has been driven and simulated. The active force generators of the designed active suspension system PID and sliding mode controllers have been introduced and designed.

The logic behind selection of this type of noise is the randomness property of it. Generally, the term noise or random fluctuations characterize all physical systems in nature. The apparently irregular or chaotic fluctuations were considered as noise in all fields except in a few such as astronomy [70]. The term "white" refers to the frequency domain characteristic of noise. Ideal white noise has equal power per unit bandwidth, which results in a flat power spectral density across the frequency range of interest. Thus, the power in the frequency range from 100 to 110 Hz is the same as the power in the frequency range from 1000 to 1010 Hz. The term "Gaussian" refers to the probability density function (pdf) of the amplitude values of a noise signal. The color of the noise refers to the frequency domain distribution of the noise signal power. Since the white noise contains all frequencies, it can be considered as random input which simulates any type of pavement condition. In this case, instead of simulating the road conditions with different color noises, simply the Gaussian white noise can be utilized [71].

In this book, MATLAB software which has been adopted in many researches [38, 72–85] has been utilized to design the controllers as well as to model the effect of the white noise on the mechanical structure and investigate the stability of the system. The results show that the designed controllers have an effective performance to eliminate the effect of road conditions which has a significant effect on reducing the fuel consumption and contributes to environment sustainability.

1.4   Methodology

Vehicle handling performance and fuel consumption rate are two important factors which are directly affected by vehicle suspension system [86]. Conventionally, to decrease the vehicle vibration, a combination set up of springs and dampers has been used. Such a set up namely is known as passive suspension system. The input disturbance to the car from pavement condition cannot be eliminated by passive suspension system since the damping ratio is constant and not adjustable. To eliminate the effect of road condition, the best solution is to utilize an autonomous control system, known as active suspension system, which can compensate for the noise input to the system by exerting force to the system [87].

In this book, problem has been defined as what the effect of the noise on quarter car model as the mechanical structure is, and how an effective controller should be designed to eliminate stochastic effect of the Gaussian white noise.

1.5   Organization of the Book

After the brief introduction, the rest of the book has been organized in following manner. First, in Chap. 2 an introduction to noise and different type of it has been given. The white noise and its applications including the mathematical modeling of white noise as pavement condition have been introduced in Chap. 3. A quarter car model and related suspension system as the mechanical structure have been introduced and modeled mathematically in Chap. 4. Next, in Chap. 5, the noise cancelation techniques with focusing on utilizing PID and sliding mode controllers have been discussed. At the end, in Chap. 6, steps of simulation set up and data collection have been described, and the proposed methodologies have been verified by analyzing the results.

References

1. M.I. Hoffert et al., Advanced technology paths to global climate stability: energy for a greenhouse planet. Science **298**(5595), 981–987 (2002)
2. P.M. Vitousek, H.A. Mooney, J. Lubchenco, J.M. Melillo, Human domination of Earth's ecosystems. Science **277**(5325), 494–499 (1997)
3. B. Dong, R.T. Sutton, A.A. Scaife, Multidecadal modulation of El Nino–Southern Oscillation (ENSO) variance by Atlantic Ocean sea surface temperatures. Geophys. Res Lett. **33**(8) 2006
4. M. Younger, H.R. Morrow-Almeida, S.M. Vindigni, A.L. Dannenberg, The built environment, climate change, and health: opportunities for co-benefits. Am. J. Prev. Med. **35**(5), 517–526 (2008)
5. S. Pacala, R. Socolow, Stabilization wedges: solving the climate problem for the next 50 years with current technologies. Science **305**(5686), 968–972 (2004)
6. S. Shafiee, E. Topal, When will fossil fuel reserves be diminished? Energy Policy **37**(1), 181–189 (2009)
7. I. Dincer, Renewable energy and sustainable development: a crucial review. Renew. Sustain. Energy Rev. **4**(2), 157–175 (2000)
8. K.A. Small, K. Van Dender, Fuel efficiency and motor vehicle travel: the declining rebound effect. Energy J., 25–51 (2007)
9. P.K. Goldberg, The effects of the corporate average fuel efficiency standards in the US. J Ind. Econ. **46**(1), 1–33 (1998)
10. R. Stone, *Motor Vehicle Fuel Economy* (Macmillan International Higher Education, 2017)
11. P. Mock, J. German, A. Bandivadekar, I. Riemersma, Discrepancies between type-approval and "real-world" fuel-consumption and CO. Int. Counc. Clean Transp. **13** (2012)
12. P. Kågeson, Reducing CO_2 emissions from new cars. Eur. Fed. Transp. Environ. (2005)
13. S. McBeath, *Competition Car Aerodynamics*, 3rd edn. (Veloce Publishing Ltd, 2017)
14. K. Holmberg, P. Andersson, N.-O. Nylund, K. Mäkelä, A. Erdemir, Global energy consumption due to friction in trucks and buses. Tribol. Int. **78**, 94–114 (2014)
15. J. Liu et al., Nanoparticle chemically end-linking elastomer network with super-low hysteresis loss for fuel-saving automobile. Nano Energy **28**, 87–96 (2016)
16. A.-H. Kakaee, P. Rahnama, A. Paykani, Influence of fuel composition on combustion and emissions characteristics of natural gas/diesel RCCI engine. J. Nat. Gas Sci. Eng. **25**, 58–65 (2015)
17. A. Dicks, D.A.J. Rand, *Fuel cell Systems Explained* (Wiley Online Library, 2018)
18. E. Khalife, M. Tabatabaei, A. Demirbas, M. Aghbashlo, Impacts of additives on performance and emission characteristics of diesel engines during steady state operation. Prog. Energy Combust. Sci. **59**, 32–78 (2017)
19. M. Zhou, H. Jin, W. Wang, A review of vehicle fuel consumption models to evaluate eco-driving and eco-routing. Transp. Res. Part D: Transp. Environ. **49**, 203–218 (2016)
20. M. Flannigan et al., Fuel moisture sensitivity to temperature and precipitation: climate change implications. Clim. Change **134**(1–2), 59–71 (2016)
21. Y. Xu, F.E. Gbologah, D.-Y. Lee, H. Liu, M.O. Rodgers, R.L. Guensler, Assessment of alternative fuel and powertrain transit bus options using real-world operations data: life-cycle fuel and emissions modeling. Appl. Energy **154**, 143–159 (2015)
22. L. Li, S. You, C. Yang, B. Yan, J. Song, Z. Chen, Driving-behavior-aware stochastic model predictive control for plug-in hybrid electric buses. Appl. Energy **162**, 868–879 (2016)

23. H. Wang, X. Zhang, M. Ouyang, Energy consumption of electric vehicles based on real-world driving patterns: a case study of Beijing. Appl. Energy **157**, 710–719 (2015)
24. S.E. Li, H. Peng, Strategies to minimize the fuel consumption of passenger cars during car-following scenarios. Proc. Inst. Mech. Eng., Part D: J. Automobile Eng. **226**(3), 419–429 (2012)
25. L. DeRaad, The influence of road surface texture on tire rolling resistance. SAE Technical Paper0148-7191, 1978
26. G. Descornet, Road-surface influence on tire rolling resistance, in *Surface characteristics of roadways: international research and technologies* (ASTM International, 1990)
27. U. Sandberg, A. Bergiers, J.A. Ejsmont, L. Goubert, R. Karlsson, M. Zöller, *Road Surface Influence on Tyre/Road Rolling Resistance* (MIRIAM, editor, 2011)
28. I. Zaabar, K. Chatti, *A Field Investigation of the Effect of Pavement Surface Conditions on Fuel Consumption* (2011)
29. F. Perrotta, L. Trupia, T. Parry, L.C. Neves, Route level analysis of road pavement surface condition and truck fleet fuel consumption, in *Pavement Life-Cycle Assessment* (CRC Press, 2017), pp. 61–68
30. N. Dhakal, M.A. Elseifi, Effects of asphalt-mixture characteristics and vehicle speed on fuel-consumption excess using finite-element modeling. J. Transp. Eng., Part A: Syst. **143**(9), 04017047 (2017)
31. M. Ziyadi, H. Ozer, S. Kang, I.L. Al-Qadi, Vehicle energy consumption and an environmental impact calculation model for the transportation infrastructure systems. J. Clean. Prod. **174**, 424–436 (2018)
32. A. Loulizi, H. Rakha, Y. Bichiou, Quantifying grade effects on vehicle fuel consumption for use in sustainable highway design. Int. J. Sustain. Transp. **12**(6), 441–451 (2018)
33. Y. Huang, E.C. Ng, J.L. Zhou, N.C. Surawski, E.F. Chan, G. Hong, Eco-driving technology for sustainable road transport: a review. Renew. Sustain. Energy Rev. **93**, 596–609 (2018)
34. M. Speckert, M. Lübke, B. Wagner, T. Anstötz, C. Haupt, Representative road selection and route planning for commercial vehicle development, in *Commercial Vehicle Technology 2018* (Springer, 2018), pp. 117–128
35. A.M. Pérez-Zuriaga, D. Llopis-Castelló, F.J. Camacho-Torregrosa, I. Belkacem, A. García, *Impact of Horizontal Geometric Design of Two-Lane Rural Roads on Vehicle CO_2 Emissions* (2017)
36. L. Liu, C. Li, X. Hua, Y. Li, Multi-factor integration based eco-driving optimization of vehicles with same driving characteristics, in *Chinese Automation Congress (CAC)* (IEEE, 2017), pp. 6871–6876
37. J. Robson, C. Dodds, Stochastic road inputs and vehicle response. Veh. Syst. Dyn. **5**(1–2), 1–13 (1976)
38. A. Azizi, Computer-based analysis of the stochastic stability of mechanical structures driven by white and colored noise. Sustainability **10**(10), 3419 (2018)
39. J. Palmer, S. Sljivar, *Vehicle Fuel Consumption Monitor and Feedback Systems* (ed. Google Patents, 2017)
40. P.J. Alvarado, Steel vs. plastics: the competition for light-vehicle fuel tanks. JOM **48**(7), 22–25 (1996)
41. Y. Kurihara, K. Nakazawa, K. Ohashi, S. Momoo, K. Numazaki, Development of multi-layer plastic fuel tanks for Nissan research vehicle-II. SAE Transa. 1239–1245 (1987)
42. G. Bahng, D. Jang, Y. Kim, M. Shin, A new technology to overcome the limits of HCCI engine through fuel modification. Appl. Therm. Eng. **98**, 810–815 (2016)
43. B. Erkuş, M.I. Karamangil, A. Sürmen, Enhancing the heavy load performance of a gasoline engine converted for LPG use by modifying the ignition timings. Appl. Therm. Eng. **85**, 188–194 (2015)
44. S. Tangöz, S.O. Akansu, N. Kahraman, Y. Malkoc, Effects of compression ratio on performance and emissions of a modified diesel engine fueled by HCNG. Int. J Hydrogen Energy **40**(44), 15374–15380 (2015)

45. M. Ben-Chaim, E. Shmerling, A. Kuperman, Analytic modeling of vehicle fuel consumption. Energies **6**(1), 117–127 (2013)
46. K. Ahn, *Microscopic Fuel Consumption and Emission Modeling* (Virginia Tech, 1998)
47. M. Ross, Automobile fuel consumption and emissions: effects of vehicle and driving characteristics. Annu. Rev. Energy Env. **19**(1), 75–112 (1994)
48. R. Smit, A. Brown, Y. Chan, Do air pollution emissions and fuel consumption models for roadways include the effects of congestion in the roadway traffic flow? Environ. Model Softw. **23**(10–11), 1262–1270 (2008)
49. H.A. Rakha, K. Ahn, K. Moran, B. Saerens, E. Van den Bulck, Virginia tech comprehensive power-based fuel consumption model: model development and testing. Transp. Res. Part D: Transp. Environ. **16**(7), 492–503 (2011)
50. T.-Q. Tang, H.-J. Huang, H.-Y. Shang, Influences of the driver's bounded rationality on micro driving behavior, fuel consumption and emissions. Transp. Res. Part D: Transp. Environ. **41**, 423–432 (2015)
51. J. Wang, H.A. Rakha, Fuel consumption model for heavy duty diesel trucks: Model development and testing. Transp. Res. Part D: Transp. Environ. **55**, 127–141 (2017)
52. Y. Wang, W. Zhao, G. Zhou, Q. Gao, C. Wang, Suspension mechanical performance and vehicle ride comfort applying a novel jounce bumper based on negative Poisson's ratio structure. Adv. Eng. Softw. **122**, 1–12 (2018)
53. W. Ren, B. Peng, J. Shen, Y. Li, Y. Yu, Study on vibration characteristics and human riding comfort of a special equipment cab. J. Sens. **2018** (2018)
54. P.B. Koganti, F.E. Udwadia, Unified approach to modeling and control of rigid multibody systems. J. Guidance, Control, Dyn. 2683–2698 (2016)
55. C.M. Pappalardo, D. Guida, Control of nonlinear vibrations using the adjoint method. Meccanica **52**(11–12), 2503–2526 (2017)
56. C.M. Pappalardo, D. Guida, Use of the adjoint method for controlling the mechanical vibrations of nonlinear systems. Machines **6**(2), 19 (2018)
57. Y. Huang, J. Na, X. Wu, X. Liu, Y. Guo, Adaptive control of nonlinear uncertain active suspension systems with prescribed performance. ISA Trans. **54**, 145–155 (2015)
58. Q. Zhu, J.-J. Ding, M.-L. Yang, LQG control based lateral active secondary and primary suspensions of high-speed train for ride quality and hunting stability. IET Control Theory Appl. **12**(10), 1497–1504 (2018)
59. J. Marzbanrad, N. Zahabi, H∞ active control of a vehicle suspension system exited by harmonic and random roads. Mech. Mech. Eng. **21**(1) (2017)
60. M.M. Elmadany, Optimal linear active suspensions with multivariable integral control. Veh. Syst. Dyn. **19**(6), 313–329 (1990)
61. H. Siswoyo, N. Mir-Nasiri, M.H. Ali, Design and development of a semi-active suspension system for a quarter car model using PI controller. J. Autom. Mobile Robot. Intell. Syst. **11** (2017)
62. H. Metered, W. Abbas, A. Emam, Optimized Proportional Integral Derivative Controller of Vehicle Active Suspension System using Genetic Algorithm. SAE Technical Paper (2018), pp. 01–1399
63. H. Li, X. Jing, H.R. Karimi, Output-feedback-based $ H_ {\infty} $ control for vehicle suspension systems with control delay. IEEE Trans. Industr. Electron. **61**(1), 436–446 (2014)
64. A.E.-N.S. Ahmed, A.S. Ali, N.M. Ghazaly, G.A. El-Jaber, PID controller of active suspension system for a quarter car model. Int J. Adv. Eng. Technol. **8**(6), 899 (2015)
65. A. Buscarino, C.F.L. Fortuna, M. Frasca, Passive and active vibrations allow self-organization in large-scale electromechanical systems. Int. J. Bifurcat. Chaos **26**(07), 1650123 (2016)
66. F. Zhao, S.S. Ge, F. Tu, Y. Qin, M. Dong, Adaptive neural network control for active suspension system with actuator saturation. IET Control Theory Appl. **10**(14), 1696–1705 (2016)
67. Y. Taskin, Y. Hacioglu, N. Yagiz, Experimental evaluation of a fuzzy logic controller on a quarter car test rig. J. Brazilian Soc. Mech. Sci. Eng. **39**(7), 2433–2445 (2017)
68. D. Singh, Modeling and control of passenger body vibrations in active quarter car system: a hybrid ANFIS PID approach. Int. J. Dyn. Control (2018)

69. M.A.Z.I.M. Fauzi et al., *Enhancing Ride Comfort of Quarter Car Semi-active Suspension System Through State-Feedback Controller* (Springer Singapore, Singapore, 2018), pp. 827–837
70. V. Marmarelis, *Analysis of physiological systems: The white-noise approach* (Springer Science & Business Media, 2012)
71. J. Hawkins Jr., S. Stevens, The masking of pure tones and of speech by white noise. J. Acoust. Soc. Am. **22**(1), 6–13 (1950)
72. A. Ashkzari, A. Azizi, Introducing genetic algorithm as an intelligent optimization technique, in *Applied Mechanics and Materials*, vol. 568. (Trans Tech Publ 2014), pp. 793–797
73. A. Azizi, Introducing a novel hybrid artificial intelligence algorithm to optimize network of industrial applications in modern manufacturing. Complexity **2017** (2017)
74. A. Azizi, Hybrid artificial intelligence optimization technique, in *Applications of Artificial Intelligence Techniques in Industry 4.0* (Springer, 2019), pp. 27–47
75. A. Azizi, Modern manufacturing, in *Applications of Artificial Intelligence Techniques in Industry 4.0* (Springer, 2019), pp. 7–17
76. A. Azizi, RFID network planning, in *Applications of Artificial Intelligence Techniques in Industry 4.0* (Springer, 2019), pp. 19–25
77. A. Azizi, *Applications of Artificial Intelligence Techniques in Industry 4.0* (ed: Springer)
78. A. Azizi, F. Entesari, K. G. Osgouie, M. Cheragh, Intelligent mobile robot navigation in an uncertain dynamic environment, in *Applied Mechanics and Materials*, vol. 367. (Trans Tech Publ, 2013), pp. 388–392
79. A. Azizi, F. Entessari, K. G. Osgouie, A. R. Rashnoodi, Introducing neural networks as a computational intelligent technique, in *Applied Mechanics and Materials*, vol. 464. (Trans Tech Publ, 2014), pp. 369–374
80. A. Azizi, N. Seifipour, Modeling of dermal wound healing-remodeling phase by Neural Networks, in *International Association of Computer Science and Information Technology-Spring Conference, 2009. IACSITSC'09*, (IEEE 2009), pp. 447–450
81. A. Azizi, A. Vatankhah Barenji, M. Hashmipour, Optimizing radio frequency identification network planning through ring probabilistic logic neurons. Adv. Mech. Eng. **8**(8), 1687814016663476 (2016)
82. A. Azizi, P. G. Yazdi, M. Hashemipour, Interactive design of storage unit utilizing virtual reality and ergonomic framework for production optimization in manufacturing industry. Int. J. Interac. Des. Manuf. (IJIDeM) 1–9 (2018)
83. M. Koopialipoor, A. Fallah, D.J. Armaghani, A. Azizi, E.T. Mohamad, Three hybrid intelligent models in estimating flyrock distance resulting from blasting. Eng. Comput. 1–14 (2018)
84. K.G. Osgouie, A. Azizi, Optimizing fuzzy logic controller for diabetes type I by genetic algorithm, in *The 2nd International Conference on Computer and Automation Engineering (ICCAE)*, vol. 2. (IEEE, 2010), pp. 4–8
85. S. Rashidnejhad, A. H. Asfia, K. G. Osgouie, A. Meghdari, A. Azizi, Optimal trajectory planning for parallel robots considering time-jerk, in *Applied Mechanics and Materials*, vol. 390. (Trans Tech Publ, 2013), pp. 471–477
86. Y. Zhang, K. Guo, D. Wang, C. Chen, X. Li, Energy conversion mechanism and regenerative potential of vehicle suspensions. Energy **119**, 961–970 (2017)
87. I. Maciejewski, T. Krzyzynski, H. Meyer, Modeling and vibration control of an active horizontal seat suspension with pneumatic muscles. J. Vib. Control 1077546318763435 (2018)

Chapter 2
Introduction to Noise and its Applications

2.1 Overview

A random fluctuation also called "noise," is a characteristic of all physical systems in nature. In most of the scientific fields, noise is considered as apparently irregular or periodic chaotic. Since 1826, noise and fluctuation have been of interest to study Brownian motion which indirectly approves molecules and atoms existence. The recognizable measured noise or signal patterns express valuable information of a system [1].

2.2 Background and History of Noise

In 1826, Robert Brown, a Scottish botanist, carried out the first research about the noise. He observed periodic motion of pollen on a water film surface [2]. The origin of Brownian motion was the first noise problem which had been unsolved for almost 80 years. In 1905, the problem was finally solved and presented by considerable work of Von Smoluchowski [2] and [3], which indirectly confirmed the molecules and atoms existence. An obvious sign of continuum molecular bombardment belongs to the medium of liquid or gas, is Brownian motion's fluctuations [4]. Nowadays, noise and fluctuation analysis are areas of interest in different scientific fields such as physical, biological, and other systems. Effects of the noise can be varying due to the circumstance. In some cases, it can enhance the system performance (constructive), while it could be destructive [5]. Nevertheless, in nonlinear systems, noise has an essential characteristic due to the possibility of the performance optimization in the levels which noise is nonzero.

© The Author(s), under exclusive license to Springer Nature Singapore Pte Ltd. 2019   13
A. Azizi and P. G. Yazdi, *Computer-Based Analysis of the Stochastic Stability of Mechanical Structures Driven by White and Colored Noise*, SpringerBriefs in Applied Sciences and Technology, https://doi.org/10.1007/978-981-13-6218-7_2

The noise can be represented with considering the following factors:

- *Distribution of duration/Size in time or space*
- *Variance or Power spectrum which shows inverse power-law* [6].

Early attempts lead to recognition of some quantitative mathematical distributions which almost fit sets of data of a vast study range and scientific disciplines. Statistical normal distribution and power law distribution are the most well-known ones.

In 1733, a gambler's consultant and statistician who was Abraham de Moivre, discovered the first normal curve of error (or the bell curve because of its shape). Normal curve importance initiated from the fact that the distributions of many natural phenomena are at least approximately normally distributed. This normal distribution concept highlights how the experimental data have been analyzed over last two centuries [6, 7].

2.3 Noise in Electronic View

In electronic knowledge, an unwanted disturbance within an electrical signal is defined as noise. The noise generation in this category is the result of many different effects. So that these types of noise greatly vary.

2.4 Noise in Communication View

Noise in an undesired disturbance which is random, and it will be considered as error in communication systems. In general, noise in communication view is all type of disturbing and unwanted energy which its resource might be man-made or natural. There is a difference between the noise and interference, and it has to be distinguished about the definition. The difference between the noise and distortion is that the distortion usually is defined as systematic unwanted alteration of a signal (wave) generated by communication devices, so it's artificial noise [8]. Also the signal itself can generate interference, for instance if there is a conflict between subsequent symbols, or not perfect matching on a transmission line. Noise is everything that is not useful signal, so can be due to interference, temperature, impurities, gamma rays, moon phase or whatever. So interference is noise but the inverse is not true.

2.5 Different Types of Noise

There are different types of noise exist based on the characteristic and sources which is generating them. Although they are different about these aspects but for various type of them some of these characteristics are common. For instance, noise spreads

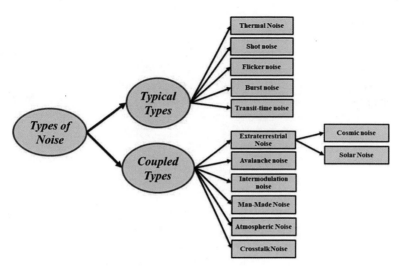

Fig. 2.1 Different types of noise categories

across the frequency spectrum, in some category the frequency are low and in some of them is high, but the amplitude of these noises might be the same (Fig. 2.1).

2.5.1 Typical Types of Noise

To categorise the different types of noise and the source which generating it, it is worthy to categorise them first based on their features. There are five main resources of noise which are common in real life and have different applications no matter as wanted or unwanted item. In order to cover all types of noises, the coupled category of noise has been distinguished from the typical types.

2.5.1.1 Thermal Noise

Molecular forces bounded many ions and electrons strongly in conductors. The vibration of ions around their average or normal position is a function of position. The source of resistance in conductors is the electrons' collision which is continuous and also the ions vibration. The collision of the electrons and vibration of ions generate continues transfer of energy between them and it is the reason of resistance in the conductors. Free electron movements create a pure random current which has average zero over the long time [9, 10].

2.5.1.2   Shot Noise

Nonlinearity in the transmitter, receiver, or intervening transmission system induces intermediation noise. These components usually react as linearly which the output is equal to the input times a constant. In a nonlinear system, the output is a more complex function of the input. The reason of nonlinearity can be the dysfunction of the component or excessive strong signal usage. Under this situation, the sum and difference terms happen [11].

2.5.1.3   Flicker Noise

Flicker noise exists in most of the electronic devices and can affect the system in different ways such as: impurities in a conductive channel, generation and recombination noise in a transistor due to base current, and many other effects. Due to the fluctuation of resistance during the transformation and fluctuation of current according to Ohm's Law, the flicker noise will be generated. As the mentioned items are the characteristics of the direct current (DC), this noise usually exists in this type of voltage. This type of noise is appearing in rotation rate of the earth, fluctuation of sand and hourglass flow and undersea currents as a low-frequency noise. The high-frequency noises are going to be considered as white noise. Nevertheless, noise with low frequency in oscillators might be diverse to frequencies close to the carrier which is going to generate oscillator phase noise. As in the all of the resistors there is carbon composition. The total noise will be increased to above thermal noise due to the flicker noise which is available in resistors, and it is named as excess noise [12].

In contrast, the least amount of flicker noise is in wire-wound resistors. Since flicker noise level is associated with DC level, at low current, thermal noise is prevailing effect in the resistor. The mentioned type of resistors might not influence the level of noise due to the frequency window [13].

2.5.1.4   Transit Time Noise

Transit time noise can be defined as the period of time which is the carrier of current. It can be considered like a hole for transport from input to output. It is important to know that the involved distances are unimportant because of the size of the devices, and the required time lasts for careering current to pass a short distance is restricted. This time is insignificant at low frequencies, whereas for the high operational frequencies the processed signal is the magnitude of the transit time. As a type of noise which is random in the system, transit time will be defined and it is proportional to the operational frequency directly [14].

2.5.1.5 Burst Noise

Burst noise is another type of electronic noise which presents in semiconductors . It is also known as random, bistable, telegraph, signal, popcorn, and impulse noise. It is immediate step-like offset among more than two current levels or discrete voltage, around hundred microvolts, at random times. Each voltage or current offset usually occurs between milliseconds to seconds and if it is connected to a speaker will sound like popcorn bursting [1].

The first witness of burst noise was in early point contact diodes, and after that it has been detected during one of the first semiconductor op-amps commercializing [15]. The single source of popcorn noise and its occurrences are not theoretically clear. However, the most frequent cause of that is charge transmit intermittent trapping and releasing in defect site in bulk semiconductor crystal or in thin film interfaces. In some circumstances where charges have a substantial effect on transistor performance, there is considerable signal output. The common cause of these defects is due to manufacturing process, for instance, heavy ion implantation or by unwanted side effects like surface contamination [16].

2.5.2 Coupled Type of Noise

With considering the mentioned description about the resources of noise within an electronic circuit, it is essential to mention the other type of noise which is couple noise. Couple noise is another noise resource which has different definition than the typical type of noise. Whenever additional noise energy is going to be coupled to the main noise resource into the electrical circuit, coupled noise will be generated. There are many types of noise energy can be coupled to the circuit, but most of them are from external environments. The following sections show different source of noise which are resulting to the coupled noise.

2.5.2.1 Atmospheric Noise

Lightning in the atmosphere induces discharges in the form of electrical disturbances , and it is known as atmospheric noise. Naturally, they are random electrical impulses. Consequently, the distribution of the energy is completely over the radio communication frequency spectrum [17]. Therefore, atmospheric noise creates fake radio signals with a wide range of frequency with distributed components. These fake radio signals consist propagated noise over the earth similar to the same frequency radio waves. Thus, a receiving antenna at a point picks up both signal and the static from all the thunderstorms, local or remote. The atmospheric noise strength nearly changes inversely with the frequency. The atmospheric noise which is very little will be generated in band of UHF and VHF and large one will be generated in broadcast bands which are low or medium bands. In addition, components of noise for VHF and UHF components are limited to less than 80 km propagation of line of sight. Therefore, at frequencies above 30 MHz, the atmospheric noise is less intense [18].

2.5.2.2   Crosstalk Noise

Crosstalk is a type of noise which might be experienced by anyone who had heard a secondary conversation while speaking on telephone. It is an undesirable signal paths coupling. It can happen by close twisted pair electrical coupling coax, cable lines and rarely, carrying multiple signals. In addition, when unsolicited signals are collected by microwave antennas, crosstalk can occur. Microwave energy distributes through the propagation despite being highly directional. Characteristically, crosstalk noise has less or same magnitude as thermal noise [19].

2.5.2.3   Interference Noise

It is the super position of two or more waves propagating through the same medium at same time without being disrupted. Depending on how maximum and minimum amplitude of the traveling waves overlapped, the resulting wave amplitude can be lower or higher compared with individual waves. When amplitudes of two interacting waves have the same sign, the amplitude of the sum wave is greater than that of the larger wave. This is known as constructive interference [20].

In contrast, when two waves interact with opposite amplitude in less than half the time, the stronger wave has bigger amplitude in comparison with the resulting wave. This interference is called destructive interference. However, it is possible that having two same waves with opposite amplitude which in destructive interference, they can cancel each other [21].

2.5.2.4   Intermodulation Noise

In a nonlinear system, including signals with two or more dissimilar frequencies, the modulation of the amplitude will be defined as intermodulation distortion or intermodulation. Additional signals will be formed between the components of each frequency during the intermodulation. These frequencies are at harmonic frequencies, summation and deference of the frequencies of the original frequencies, and also multiples of all of the mentioned frequencies [22, 23].

By considering the characteristic of the signal processing which probably is nonlinear, intermodulation can be defined as the result. In order to calculate the nonlinearity, Volterra series and Taylor series can be utilized [24].

Since intermodulation causes unfavorable spurious emissions usually in the form of sidebands, it is seldom favorable in radio or audio processing. Intermodulation augments the occupied bandwidth for radio transmissions causing interference in adjacent channel. Therefore, the range of the spectrum can be increased, and the quality of the audio can be reduced. The definition of the intermodulation is different than the musical application's harmonic distortion or intentional modulation in applications in which modulated signals show up as intentional nonlinear element.

The products of intermodulation in audio is related to the input frequency and they are not harmonic [24].

2.5.2.5 Man-Made Noise (Industrial Noise)

Another category of noises is man-made noise or industrial noise. Some examples of this type of noises are: electrical noise generated by ignition in aircrafts, vehicles with normal and electrical motors, different type of electrical machines which are heavy, fluorescent lights, and lines of high voltage. The noises in this category are generated in the mentioned systems and devised by the arc discharge while they are functioning. High intensity of these types of noises is localized in industrial and highly areas. The intensity of the noises from these areas is higher than all other sources and their frequency ranged between 1 and 600 MHz [25–27].

2.5.2.6 Extra-Terrestrial or Galactic Noise

The definition of this type of noise refers to the type of radio noise which its resource is out of the earth from the galaxy. This type of noise is known as galactical noise, and can be presented in some degree with all of the possible directions. It is important to know that it is mostly penetrating near to the galaxy plane and mainly in galactic center direction. This type of noise has a wide range of frequency between 1 and 1360 mc. The Extra-terrestrial noise mainly can be categorized to the solar noise and cosmic noise [28, 29].

Solar Noise

Solar noise is a type of electrical noise which its source is the sun. As the surface of the sun has the temperature of over 6000 °C, and it is a large object with the steady situation, it generates the steady radiation of noise. This noise is actually electrical energy which is radiating with the frequency which has a wide range. The solar noise spectrum is the same spectrum in radio communication. The noise produced by the sun has a time varying intensity. This noise cycle is 11 years. The noise generated by sun at the peak of the cycle induces great interference to radio signal, causing many unusable for communications. During the rest of cycle, the noises are at a minimum level [30, 31].

Cosmic Noise

Distant stars also have high temperature and emit noises like sun. The thermal noises (or black body noise) received from distant stars are distributed almost homogeneously over sky. As the other type of sources which can generate this type of noise, distant galaxies, centre of Milky Way and pulsars and quasars as virtual point sources can be mentioned [32, 33].

2.5.2.7   Avalanche Noise

At the beginning of the avalanche breakdown during a junction diode functioning, avalanche noise will be generated. The reason behind this noise generation is a semiconductor junction phenomenon. It carriers the voltage gradient which is high and developing adequate energy to dislodge additional carriers through physical impact [34].

2.5.2.8   Impulse Noise

This type of noise is defined as minor disturbance for analog data. As an example, crackles or short clicks can disrupt voice transmission with no effect on its intelligibility. In digital data communication, this type of noise is the main reason of errors [35].

2.5.3   Colored Noise in Signal Processing View

Noise in signal processing has different way of definition. In signal processing, noise will be categorized based on its statistical properties. These statistical properties are defined as noise color. It is difficult to differentiate the various types of noise. However, it is possible to differentiate the noises with color method. The noise colors arise from similarity to light and correspond to the content of frequency. For each noise, a color is defined in which some of them are related to the physical world and some are adopted for psychoacoustics field. These colors resemble the frequency power which is proportional to their spectrum see (Table 2.1) [15].

Table 2.1 Colored noise and frequencies [15]

Color	Frequency content
Purple/Violet	f^2
Blue	f
White	1
Pink	$1/f$
Red/Brown	$1/f^2$

2.5.3.1 Purple or Violet Noise

Purple noise is also called violet noise with increasing power density of 6 dB per octave with increasing frequency (frequency content of f 2) over the range of finite frequency. Since it is being the result of white noise signal differentiation, it is also known as differentiated white noise [36].

2.5.3.2 Blue Noise

Blue Noise is considered as white noise with high frequency. The spectral density of this noise color is proportional to its frequency. As the blue color is the higher end of visible light frequency spectrum, this noise also named as azure noise which comes from optics. This means that as the frequency increases, the signal energy and power (density proportional to f) will increase.

In blue noise, each octave continuously increases by 3 dB. It is one of the unique characteristics of this noise which packs same amount of energy in each octave as it is in the combination of the amount of energy in two octaves below it [37, 38].

2.5.3.3 Pink Noise

In cases in which the spectral density power is proportional and in reverse direction of the signal frequency, the frequency spectrum belongs to pink noise which also known as 1/f noise. The amount of energy which is carried by each octave is equal in pink noise. The name pink noise is because of its power spectrum which appears in visible light. The pink noise is in contrast with white noise because of the equal intensity per frequency interval in white noise [39].

This type of noise is useful due to the same frequency which is hearable for human. Pink noise will be utilized as useful signal for testing the audio devices such as speakers and amplifiers. As the real example of pink noise, frequency of heart beats and activity of neural and DNA statistical sequences are worth to be mentioned [40].

2.5.3.4 Red/Brown Noise

As one of the colored noises, brown noise includes some other noises such as pink, white, and blue noises. This noise is named as brown because of the Robert Brown due to his research about the Brownian motion as the random particle motion. So that, because of the mentioned characteristic which is changing the sound signal during the time randomly, it is also known as Brownian noise.

This type of noise has a spectral density which is proportional to its frequency and contrariwise. This characteristic is completely in opposite definition of white noise which has even spectral density for all of the frequencies. It means brown noise power obviously decreases with increase in the frequency.

Therefore, at lower frequencies, brown noise has considerably higher energy compared with its higher frequencies. As the red light has the low frequency, the brown noise will be named as red noise in some cases. Brown and white noises are similar for our ear. The brown noise is much deeper, and it is like a roar but low. However, white noise is not this much deep and it is like a strong waterfall sound.

2.5.3.5   White Noise

In the middle of a spectrum which runs from purple trough red/brown is white noise. The term "white" corresponds to noise frequency domain characteristic of. Ideal white noise has equal power per unit bandwidth, which results in a flat power spectral density across the frequency range of interest. Therefore, the power in the frequency range from 100 to 110 Hz is the same as the power in the frequency range from 1000 to 1010 Hz [41]. It is not possible to achieve infinite bandwidth due to the infinite amount of required power when signal of white noise is generated in actual life. Nevertheless, it is possible to create signals of white noise over the desired frequencies. For a specified frequency range, power per hertz is equal for a uniform white noise. The noise color is associated with the distribution of the frequency domain of the noise signal power [42].

As the white noise is the main focused category of noise in this book, in the next chapter the white noise and its applications and different types of mathematical modeling belong to this noise and its applications is described.

References

1. A.M. Selvam, Noise or random fluctuations in physical systems: a review, in *Self-organized Criticality and Predictability in Atmospheric Flows* (Springer, 2017), pp. 41–74
2. M. Von Smoluchowski, Ann. Phys. **326**, 756 (1906)
3. A. Einstein, Investigations on the theory of the Brownian movement. Ann. der Physik (1905)
4. P. Dayan, L.F. Abbott, *Theoretical Neuroscience* (MIT Press, Cambridge, MA, 2001)
5. D. Abbott, Overview: unsolved problems of noise and fluctuations. Chaos: Interdisc. J. Nonlinear Sci. **11**(3), 526–538 (2001)
6. W.H. Press, Flicker noises in astronomy and elsewhere. Comments Astrophys. **7**, 103–119 (1978)
7. E.W. Montroll, M.F. Shlesinger, *On the Wonderful World of Random Walks* (1984)
8. H.E. Rowe, *Signals and Noise in Communication Systems* (1965)
9. J. Fields et al., Guidelines for reporting core information from community noise reaction surveys. J. Sound Vib. **206**(5), 685–695 (1997)
10. P.R. Saulson, Thermal noise in mechanical experiments. Phys. Rev. D **42**(8), 2437 (1990)
11. Y.M. Blanter, M. Büttiker, Shot noise in mesoscopic conductors. Phys. Rep. **336**(1–2), 1–166 (2000)
12. A. Szewczyk, J.S. Lentka, P. Babuchowska, F. Béguin, in *Measurements of Flicker Noise in Supercapacitor Cells*. International Conference on Noise Fluctuations, ICNF, 2017, vol. 2017, pp. 2–5

13. K. Ioka, *Flicker Noise Detection Apparatus, Flicker Noise Detection Method, and Computer-Readable Storage Device Storing Flicker Noise Detection Program* (ed: Google Patents, 2014)
14. M. Trippe, G. Bosman, A. Van Der Ziel, Transit-time effects in the noise of Schottky-barrier diodes. IEEE Trans. Microw. Theory Tech. **34**(11), 1183–1192 (1986)
15. B. Carter, R. Mancini, *Op Amps for Everyone* (Newnes, 2017)
16. Z.Y. Chong, W.M. Sansen, *Low-Noise Wide-Band amplifiers in Bipolar and CMOS Technologies* (Springer Science & Business Media, 2013)
17. A. Watt, E. Maxwell, Characteristics of atmospheric noise from 1 to 100 kc. Proc. IRE **45**(6), 787–794 (1957)
18. M. Lisi, C. Filizzola, N. Genzano, R. Paciello, N. Pergola, V. Tramutoli, Reducing atmospheric noise in RST analysis of TIR satellite radiances for earthquakes prone areas satellite monitoring. Phys. Chem. Earth, Parts A/B/C **85**, 87–97 (2015)
19. I. Catt, Crosstalk (noise) in digital systems. IEEE Trans. Electron. Comput. **6**, 743–763 (1967)
20. J.C. Reynolds, Syntactic control of interference, in *Algol-like Languages* (Springer, 1997), pp. 273–286
21. A.H. Dictionary, *The American Heritage Science Dictionary* (Houghton Mifflin Company, 2005)
22. K. Chang, Intermodulation noise and products due to frequency-dependent nonlinearities in CATV systems. IEEE Trans. Commun. **23**(1), 142–155 (1975)
23. K. Sarrigeorgidis, T. Tabet, S.A. Mujtaba, *Intermodulation Cancellation of Third-Order Distortion in an FDD Receiver* (ed: Google Patents, 2016)
24. G. Breed, Intermodulation distortion performance and measurement issues. High Freq. Electron. **2**(05), 56–57 (2003)
25. E.N. Skomal, *Man-Made Radio Noise* (Van Nostrand Reinhold Co., New York, 1978), 347 p.
26. D. Middleton, Man-made noise in urban environments and transportation systems: models and measurements. IEEE Trans. Commun. **21**(11), 1232–1241 (1973)
27. A.N. Popper, A. Hawkins, *The Effects of Noise on Aquatic Life II* (Springer, 2016)
28. A.G. Smith, Extraterrestrial noise as a factor in space communications. Proc. IRE **48**(4), 593–599 (1960)
29. H. Ko, The distribution of cosmic radio background radiation. Proc. IRE **46**(1), 208–215 (1958)
30. E.O. Elgaroy, *Solar Noise Storms: International Series in Natural Philosophy* (Elsevier, 2016)
31. R. Payne-Scott, D. Yabsley, J. Bolton, Relative times of arrival of bursts of solar noise on different radio frequencies. Nature **160**(4060), 256 (1947)
32. W.T. Sullivan III, *Cosmic Noise: A History of Early Radio Astronomy* (2009)
33. M. Kundu, F. Haddock, A relation between solar radio emission and polar cap absorption of cosmic noise. Nature **186** (1960)
34. H.K. Gummel, J.L. Blue, A small-signal theory of avalanche noise in IMPATT diodes. IEEE Trans. Electron. Devices **14**(9), 569–580 (1967)
35. R. Garnett, T. Huegerich, C. Chui, W. He, A universal noise removal algorithm with an impulse detector. IEEE Trans. Image Process. **14**(11), 1747–1754 (2005)
36. A. Raghib, B.A. El Majd, B. Aghezzaf, An Optimal deployment of readers for RFID network planning using NSGA-II, in *Recent Developments in Metaheuristics* (Springer, 2018), pp. 463–476
37. R.A. Ulichney, Dithering with blue noise. Proc. IEEE **76**(1), 56–79 (1988)
38. J. Castro, *What Is Blue Noise?* https://www.livescience.com/38583-what-is-blue-noise.html (2013)
39. P. Szendro, G. Vincze, A. Szasz, Pink-noise behaviour of biosystems. Eur. Biophys. J. **30**(3), 227–231 (2001)
40. G. Vasilescu, *Electronic Noise and Interfering Signals: Principles and Applications* (Springer Science & Business Media, 2006)
41. A. Azizi, Computer-based analysis of the stochastic stability of mechanical structures driven by white and colored noise. Sustainability **10**(10), 3419 (2018)
42. National-Instruments, *An Introduction to Noise Signals*. http://www.ni.com/white-paper/3006/en/#toc4

Chapter 3
White Noise: Applications and Mathematical Modeling

3.1 Overview

A random signal with different frequency but with an equal intensity is defined as white noise [1]. This feature gives white noise a spectral density power which is constant [2]. The mentioned definition has similar aspects in different technical and scientific fields such as telecommunications, acoustic, physics, and engineering. White noise is going to be a signal source and is utilized for statistical model signal.

3.2 Practical and Real-Life Applications of White Noise

This section tries to describe the application of white noise. Due to the behavior of white noise and its features, it has different application with different definitions. In some cases, this type of noise has advantages of utilization and, on the other hand, in some cases it is going to be a type of disturbance and disadvantageous parameter. Here, some of the well-known applications are described.

3.2.1 White Noise in Electronics Engineering

In electrical circuits, in order to achieve the impulse response, white noise can be utilized specifically in audio device and in amplifiers. White noise is not applicable to test the devices such as loudspeakers due to the great spectrum and the high frequency. As it mentioned in Chap. 2, pink noise is applicable for such conditions due to its equal amount of energy in each octave, which is different from white noise [3].

© The Author(s), under exclusive license to Springer Nature Singapore Pte Ltd. 2019   25
A. Azizi and P. G. Yazdi, *Computer-Based Analysis of the Stochastic Stability of Mechanical Structures Driven by White and Colored Noise*, SpringerBriefs in Applied Sciences and Technology, https://doi.org/10.1007/978-981-13-6218-7_3

3.2.2   White Noise in Building Acoustics

The application of white noise in acoustic is limited to cute or boost of frequency in the room or building. In order to have the best equalization setup in a room or building such as venue, by utilizing a public address (PA) system, a burst of white noise with be send to different points of the building. In this case, the engineer can evaluate that acoustic of the building about cutting or boosting any frequency [4].

3.2.3   White Noise for Tinnitus Treatment

In the sound masking technology and especially for tinnitus maskers, white noise will be utilized as the most common artificial noise source. The devices which are generating white noise (white noise machines) are available in the market as the devices for enhancing the privacy and also as sleep aids to mask the tinnitus. The most available, easy access and cheap source, and untuned FM radio frequency can generate the white noise. It is noticeable to mention that the generated white noise by this condition is so weak in comparison with other sources of white noise [5].

3.2.4   White Noise to Improve the Work Environment

There are different ideas about the effect of white noise on working environment. Generally, the effect of white noise has a direct relationship with the attention-deficit/hyperactivity disorder (ADHD) and cognitive functioning of the people in working environment. Researchers believe that white noise enhances the cognitive functioning in the people with ADHD, but it has opposite effect on the people without ADHD [6].

3.2.5   White Noise in Earthquake Simulation

In the case of earthquake engineering, non-stationary signals take the essential place. Fourier analyzing as the common method is disabled to describe the characteristic of non-stationary processes in the case of evolutionary spectral and time dependency. It needs an additional tool which gives the ability of localization of frequency and time elsewhere than the Fourier analysis. In the case of earthquake analysis, the signal which is non-stationary, and its spectral analysis do not have the description ability of the features of local transient because of the duration of the signal is above the average. In this case, utilizing white noise will help to overcome the difficulties of the analyzing. The function of the impulse response of the system and the response of the system (linear) to white noise which is stationary can be described in an identical spectral. On the other hand, both of them have different history of timing. The one that has white noise can be characterized by white noise filtration and it is the main subject has been focused in this section [7]. In order to clarify the mentioned state, the modeling of earthquake for mechanical structures with the aid of white noise has been explained in the following section.

3.2.5.1 White Noise for Earthquake Modeling for Mechanical Structures

In several researches, white noise has been used for modeling of the earthquake in mechanical structures such as buildings and offshore structures [8]. The frequency time variation can be achieved by dividing the domain of time into three continuous regions. It is needed for the ground motion to be simulated in each region. This simulation should be able to process a function of power spectral density which is unimodal and unique [9].

The stochastic process of the earthquake $\{a(r)\}$ can be simulated by this model. It is possible to reach to this goal by contiguous region modulating of white noise. This modulating should be done on a Gaussian white noise which is filtered and is utilizing a deterministic time envelope function. Each sample function of the process that can be assumed as artificial earthquake can be modeled mathematically as shown in Eqs. (3.1)–(3.21) [7–9]:

$$^{(k)}a(t) = \psi(t) \sum_{i=1}^{3} \Lambda_i(t)^{(k)} S_i(t) \tag{3.1}$$

Where:

$^{(k)}a(t)$ = kth artificial earthquake of the ensemble
$^{(k)}S_i(t)$ = kth filtered white noise sample function for the ith time region
$\Lambda_i(t)$ = deterministic rectangular time window $\equiv U(t - t_{i-1}) - V(t - t_i)$
$\psi(t)$ = deterministic envelope or shaping function.

Consider the fact that each contiguous stochastic process $\{S_i(t)\}$ is able to do the processing of a function of a unique power spectral density. By sample function $^{(k)}n(t)$ filtration out of the stochastic process of a nonzero mean Gaussian white noise $\{n(t)\}$, The sample function $^{(k)}S_i(t)$ of the stochastic process $\{S_i(t)\}$ will be achieved. Here, the way of developing is described.

Consider a zero-mean value function of a white noise process $\{n(t)\}$ which has,

$$E\langle n(t)\rangle = 0 \tag{3.2}$$

and also it has a function which is autocorrelation

$$R_{nn}(t_1, t_2) = E \langle n(t_1)n(t_2) \rangle \tag{3.3}$$

$$= R_n(\tau) = W_0 \delta(\tau) \tag{3.4}$$

Here:

$\tau \equiv t_2 - t_1$,
W_0 is a positive real constant and,
$\delta(\tau)$ is the Kronecker delta.

The samples $^{(k)}S_i(t)$ which are the sample functions of the filtered Gaussian white noise process can be achieved by

$$^{(k)}S_i(t) = \int_{-\infty}^{+\infty} h_i(t - \tau_1)^{(k)}n(\tau_1)d\tau_1; \quad -\infty \leq t \leq \infty \tag{3.5}$$

Here:

$h_i(r)$ is the impulse response function.

By considering the definition of the expected mean square acceleration and Eqs. (3.1)–(3.5), we can have

$$E\langle a^2(t) \rangle = W_0 \psi^2(t) \sum_i^3 \sum_j^3 \Lambda_i(t)\Lambda_j(t) \int_{-\infty}^{+\infty} h_i(t - \tau_1)h_j(t - \tau_1)d\tau_1 \tag{3.6}$$

and also

$$\Lambda_i(t)\Lambda_j(t) = \begin{cases} \Lambda_i(t); & i = j \\ 0; & i \neq j \end{cases} \tag{3.7}$$

We can rewrite the mean square value function as

$$E\langle a^2(t) \rangle = W_0 \psi^2(t) \sum_i^3 \Lambda_i(t) \int_{-\infty}^{+\infty} h_i^2(t - \tau_1)d\tau_1 \tag{3.8}$$

There is the fact that says

$$E\langle S_i^2(t) \rangle = W_0 \int_{-\infty}^{+\infty} h_i^2(t - \tau_1)d\tau_1 \tag{3.9}$$

By normalizing the $h_i(\tau)$, we have

$$E\langle S_i^2(t)\rangle = 1 \tag{3.10}$$

And finally we can get

$$E\langle a^2(t)\rangle = \psi^2(t) \tag{3.11}$$

It is recognizable that acceleration against the representation of time history can represent only a stochastic process's sample function because the ground acceleration is a stochastic process. So that based on the previous research for mean square acceleration, a mathematical representation can be considered which shows:

$$E\langle a^2(t)\rangle = \beta t^\gamma e^{-\alpha t} \tag{3.12}$$

(3.12) shows that the mean square acceleration achieved from superposition of wave pulses which are non-stationary and a large number.
Here:

β is an intensity parameter
α and γ are the envelope shape characterization parameters.

By considering the (3.11) and its deterministic envelope function $\psi(t)$ and utilizing (3.12), we can get the $\psi(t)$ as:

$$\psi(t) = t^{0.5\gamma} e^{-0.5\alpha t} \sqrt{\beta} \tag{3.13}$$

In order to satisfy the (3.10), there should be a function with uni-modal power-spectral density for each of the stochastic process of ground acceleration's time regions

$$S_i(\omega) = S_0 |\omega|^P e^{-|\omega Q|}; \quad -\infty \le \omega \le \infty \tag{3.14}$$

(3.14) is applicable if the value of P, S_0, and Q for each time region has a unique value. In this step, the simulation of $\{S_i(t)\}$ which is the filtered white noise process must be done. $h_i(\tau)$ is essential to be determined to follow such the simulation. Corresponding to (3.14), $h_i(\tau)$ is the parameter of the impulse response function.
The $h_i(\tau)$ function is presented as

$$h_i(\tau) = \frac{1}{2\pi} \int_{-\infty}^{+\infty} H_i(\omega) e^{j\omega\tau} d\omega \tag{3.15}$$

Here, $H_i(\omega)$ can be defined as the response function which has the complex frequency. This function is depending on a time invariant filter/system's input and output. The polar representation of the function $H_i(\omega)$ is as follows:

$$H_i(\omega) = |H_i(\omega)|e^{-j\phi i(\omega)} \tag{3.16}$$

Here, the phase and module of the $H_i(\omega)$ are shown as $\phi_i(\omega)$ and $|H_i(\omega)|$.

Focusing on (3.15) and (3.16) shows that determining the parameters which state $|H_i(\omega)|$ and $\phi_i(\omega)$ will lead to determining the $h_i(\tau)$.

When the power spectral function of $\{S(t)\}$ is as (3.16), the modulus $|H_i(\omega)|$ will be obtained.

$$W_0|H_i(\omega)|^2 \tag{3.17}$$

By utilizing (3.14) and (3.17) for each time region, we have

$$W_0|H_i(\omega)| = \sqrt{(\pi S_0|\omega|^{P/2}e^{-0.5|\omega|Q}}; \quad -\infty \leq \omega \leq +\infty \tag{3.18}$$

Due to the fact that, for determining the $\phi_i(\omega)$, there is not any available condition which is unique, any filter or system including a $|H_i(\omega)|$ which can satisfy (3.18) can be used. For this step also, it is possible to use $\phi_i(\omega)$ which is an arbitrary to perform the simulation.

Additional arbitrary introduction makes us always able to describe the filter. Imposing the arbitrary condition as it is in (3.19) makes the computations more convenient.

$$\phi_i(\omega) = 0 \tag{3.19}$$

If (3.19) is imposed, then $h_i(\tau)$ is an even function of τ and Eqs. (3.15) and (3.18) lead to a filter whose impulse response function is given by

$$h_i(\tau) = \left(\frac{S_0}{\pi}\right)\left[\frac{\Gamma(0.5P + 1)}{(Q^2/4 + t^2)^{(P+2)/4}}\right]\cos\left[(\frac{P}{2} + 1)\theta\right] \tag{3.20}$$

where

$$\theta = \tan^{-1}\left[\frac{2\tau}{Q}\right] \tag{3.21}$$

$\Gamma(.)$ is the gamma function.

P, S_0, and Q should be defined to have the complete impulse response function establishment. In addition, due to the $h_i(\tau) \neq 0$ for $\tau < 0$ fact impulse response function is not causal. The mentioned type of filter shows that if the input is zero for $t < 0$, the output may in general be nonzero for $t < 0$. Due to this statement, these filters are not realizable in physical point of view. However, Non- realizable filters for artificial simulation of earthquakes are inappropriate.

In the section above it has been tried to expose the artificial earthquake simulation, characterization parameters and numerical values which must be selected. Specifically, value of α, β and, γ parameters as the values of function of three-time modulating envelope should be selected. For next step, numerical values of P and Q for each region of contiguous time should be defined. In addition, the number of utilizing time regions and time duration of each region should be considered.

3.2.5.2 Wavelet Method Mathematical Modeling of Gaussian White Noise for Wind and Earthquake

During the past two decays, artificial wave generation has been developed in laboratories rapidly. The reason behind this development is the benefits of it in computer hardware and control system theories. Researches focusing on generating the laboratory made waves which has the nature of the real wave trains [10]. The frequency and time make the "wavelet" an appropriate tool for non-stationary signals simulation. There are two ways to accomplish this task which are a target spectrum and modulator function or giving a parent non-stationary signal, or for each octave. The Eqs. (3.22)–(3.25) [10] are utilized to clarify the mentioned states.

Consider a local wind velocity record. (3.22) shows the statistical simulation is parent process similarly.

$$\hat{x}(n) = IWT(w(n) * DWT(x(n))) \tag{3.22}$$

Here:

discrete wavelet transform (DWT), is the parent signal,
Gaussian white noise of unit variance $w(n)$,
The inverse wavelet transform (IWT).

In order to continue the definition, it is essential to mention modulated stationary process. This concept has been used by several researchers as a process to model the ground motions [9, 11, 12]. The process is cantered at frequencies which are narrow banded. Here, different modulating functions have been utilized to modulate components process.

$$x(t) = \sum_j m_j(t)S_j(t) \tag{3.23}$$

Here:

m_j is jth modulator
S_j is jth stationary component process (constant over a frequency band)
$x(t)$ is combination of different approaches to modeling m_j and S_j.

For the next step, the modulator should be normalized as following:

$$\int_{-\infty}^{\infty} m_j^2(t)dt = 1 \tag{3.24}$$

In order to perform such modeling, DWT will provide an appropriate solution [12, 13]. The parent signal's modulator function will be achieved as follows:

$$m_{ij} = A_i \sqrt{2^{i+2-M}} \frac{|a_{ij}|}{\sqrt{S_i}} \tag{3.25}$$

Here:

a_{ij} is the measured coefficients of wavelet
A_i is a level-dependent amplitude constant
S_i is the energy corresponding to the ith octave from the power spectrum.

By the aid of DWT, a ground motion has been divided into octave bands. An example of this ground motion can be earthquake. Along with the resulting modulator function, the filtered time histories of several bands can be defined.

As the modulator functions and target spectrum are given, by finding each octave energy from the target spectrum the simulation will be done. The simulation process's wavelet coefficients are multiplied with the appropriate modulator and by normalizing them in the way that the energy will be equal to the corresponding octave. After all, the coefficients which are modulated and normalized will be multiplied by white noise and inverse wavelet transformed [13]. In whole, the wavelet method explanation earthquake was the main application. All of the mathematical modeling and assumptions were performed for earthquake simulation. As long as the non-stationary signals were the target of the wavelet method, wave and wind can be simulated in the same manner due to the non-stationary behavior of them. As the proof of this statement, an example of measured non-stationary wind velocity can be beneficial.

3.2.6 Mathematical Modeling of Gaussian White Noise as Pavement Condition

In this section, to simulation of different condition of road, white noise is utilized as the types of road input signal. The advantage of considering the white noise as the road input signal in simulation is the importance of vehicle suspension system simulation. In order to have a successful vehicle suspension simulation, it is essential to have an accurate road simulation which shows the road condition during the car driving on it. Before describing this simulation, it is worthy to mention that the vehicle is assumed as a linear system [14].

3.2.6.1   White Noise Road Input Signal

Based on reviewing many literatures, it is nearly impossible to describe the road roughness with mathematical relations accurately when the speed of the vehicle is constant on the road. The reason behind that is stochastic behavior of this process which is subjected to Gauss distribution. As white noise has the characteristics which are statistical, and its definition is corresponding with the vehicle speed power spectral density, it easily can be transformed to the model of time domain of road roughness. In order to implement this simulation idea, MATLAB software can be utilized as an appropriate platform.

The actual pavement condition can be simulated properly by the white noise road input signal transformation. Whenever for the vibration input of road roughness for vehicle the actual pavement condition is utilized, it is including random character. This random character is always used to describe the statistical properties of the road power spectral density.

(3.26) shows the fitting expression of the road surface vertical displacement power spectral density (PSD). It is based on the ISO/TC108/SC2N67 international standard and presented as following [15]:

$$G_q(n) = G_q(n_0)(\frac{n}{n_0})^{-W} \tag{3.26}$$

Here:

n is Spatial frequency, the reciprocal of the wavelength, unit m^{-1}
n_0 is Reference Spatial frequency, $n_0 = 0.1 \ m^{-1}$
$G_q(n_0)$ is Road roughness coefficient, unit m^2/m^{-1}
W is the index of the Frequency, (structure of road surface frequency spectrum and usually $W = 2$).

It is noticeable to mention that when the road signal is an input, the velocity is considerable. Based on the vehicle velocity (v), the $G_q(n)$ which is spatial-frequency power spectral density is transferable to $G_q(f)$ which is time-frequency power spectral density. (3.27) to (3.30) show that the time frequency is the product of n and v when a vehicle drives through a section of road roughness [16].

$$f = v \times n \tag{3.27}$$

Here:

f is time frequency with unit of (s^{-1})
n is spatial frequency unit of (m^{-1})
v is vehicle velocity is unit of (ms^{-1}).

The mathematical relationship between $G_q(f)$ and $G_q(n)$ can be described as:

$$G_q(f) = \frac{1}{v}G_q(n) \tag{3.28}$$

(3.28) can be rewritten as following when $W = 2$

$$G_q(f) = \frac{1}{v}G_q(n_0)(\frac{n}{n_0})^{-2} = G_q(n_0)n_0^2\frac{v}{f^2} \qquad (3.29)$$

By determining the derivation of (3.29), the vehicle vertical speed power spectral density can be obtained as:

$$G_{\dot{q}}(f) = (2\pi f)^2 G_q(f) = G_q(n_0)n_0^2\frac{v}{f^2} \qquad (3.30)$$

In order to generate the model of road elevation time domain as steady Gaussian distribution stochastic process, there are several solutions. These solutions are random sequence generation method, filtering superposition method, fast Fourier inverse transform generation method (IFFT), and filtering white noise generation method. Here, filtering white noise generation method has been selected, due to its characters which are easy computing and clear physical meaning. See (3.31) [17]:

$$\dot{q}(t) = -2\pi f_0 q(t) + 2\pi n_0\sqrt{G_0(n_0)v}.w(t) \qquad (3.31)$$

Where:

$\dot{q}(t)$ is random road input signal
f_0 is filter lower-cut-off frequency
$G_q(n_0)$ is road roughness coefficient with unit of (m^2/m^{-1})
$w(t)$ is Gaussian white noise.

By conducting of Laplace transformation on (3.31), we can get the following [18]:

$$\frac{q(s)}{w(s)} = \frac{2\pi n_0\sqrt{G_q(n_0)v}}{s + 2\pi f_0} \qquad (3.32)$$

By considering the $f_0 = 0$, it will become in integrator form.

To clarify the mentioned equations, an example is provided about a vehicle with speed of 20 m/s, class road has been selected as C type (Table 3.1), $G_q(n_0)$ is equal to $256 * 10^{-6}$, then the signal of road roughness input is as it shows in Fig. 3.2 is based on the Simulink model which is illustrated in Fig. 3.1.

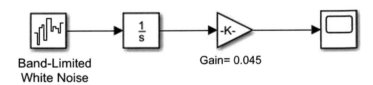

Band-Limited
White Noise

Gain= 0.045

Fig. 3.1 Simulink model of the signal of White noise pavement roughness

Table 3.1 Road roughness standard [19]

Road level	$G_q(n_0)/(10^{-6}\,\mathrm{m}^3)$ $(n_0 = 0.1\,\mathrm{m}^{-1})$	$\sigma_q/(10^{-3}\,\mathrm{m})$ $0.011\,\mathrm{m}^{-1} < n < 2.83\,\mathrm{m}^{-1}$
A	16	3.81
B	64	7.61
C	256	15.23
D	1024	30.45
E	4096	60.90
F	16,384	121.80
G	65,536	243.61
H	262,144	487.22

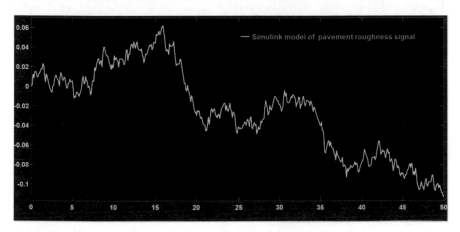

Fig. 3.2 Simulink model of Class C pavement roughness signal

The road level which has been mentioned in last example is based on an international standard, which has eight road levels A–H, which A shows the best level and H shows the worst one. The following table shows this standard value for different road level.

In order to clarify the mentioned definition and simulation methods about the white noise and its application on modeling the non-stationary models such as road excitation which has been explained in this chapter, a specific application will be focused in next chapter which is car suspension system. As the mandatory knowledge of suspension system and monitoring of pavement method, the mechanical structure should be clearly defined. In addition, the way that mechanical structure (specifically suspension system) can be modeled mathematically should be explained as well. The following chapter covered the mentioned items in detail.

References

1. A. Azizi, Computer-based analysis of the stochastic stability of mechanical structures driven by white and colored noise. Sustainability **10**(10), 3419 (2018)
2. B. Carter, R. Bruce, *Op Amps for Everyone, Texas Instruments*, ISBN, vol. 80949487 (2009), pp. 10–11
3. S.V. Vaseghi, *Advanced Digital Signal Processing and Noise Reduction* (Wiley, 2008)
4. G.-B. Stan, J.-J. Embrechts, D. Archambeau, Comparison of different impulse response measurement techniques. J. Audio. Eng. Soc. **50**(4), 249–262 (2002)
5. G. Andersson, Tinnitus loudness matchings in relation to annoyance and grading of severity. Auris Nasus Larynx **30**(2), 129–133 (2003)
6. G.B. Söderlund, S. Sikström, J.M. Loftesnes, E.J. Sonuga-Barke, The effects of background white noise on memory performance in inattentive school children. Behav. Brain Funct. **6**(1), 55 (2010)
7. K. Gurley, A. Kareem, Applications of wavelet transforms in earthquake, wind and ocean engineering. Eng. Struct. **21**(2), 149–167 (1999)
8. P. Ruiz, J. Penzien, Stochastic seismic response of structures. J. Eng. Mech. Div. **97**(2), 441–456 (1971)
9. G. Rodolfo Saragoni, G.C. Hart, Simulation of artificial earthquakes. Earthquake Eng. Struct. Dynam. **2**(3), 249–267 (1973)
10. M.J. Ketabdari, A. Ranginkaman, Numerical simulation of random irregular waves for wave generation in laboratory flumes. AUT J. Model. Simul. **43**(1), 1–6 (2011)
11. H. Kameda, Evolutionary spectra of seismogram by multifilter. J. Eng. Mech. Div. **101**(6), 787–801 (1975)
12. Y. Li, A. Kareem, Simulation of multivariate nonstationary random processes by FFT. J. Eng. Mech. **117**(5), 1037–1058 (1991)
13. K. Gurley, A. Kareem, On the analysis and simulation of random processes utilizing higher order spectra and wavelet transforms, in *Proceedings of the 2nd International Conference on Computational Stochastic Mechanics* (1994)
14. J. Sun, Y. Sun, Comparative study on control strategy of active suspension system, in *2011 Third International Conference on Measuring Technology and Mechatronics Automation (ICMTMA)*, vol. 1 (IEEE, 2011), pp. 729–732
15. J. Cao, H. Liu, P. Li, D. Brown, Adaptive fuzzy logic controller for vehicle active suspensions with interval type-2 fuzzy membership functions, in *IEEE International Conference on Fuzzy Systems, 2008. FUZZ-IEEE 2008 (IEEE World Congress on Computational Intelligence)* (IEEE, 2008), pp. 83–89
16. L. Menard, Y. Broise, Theoretical and practical aspect of dynamic consolidation. Geotechnique **25**(1), 3–18 (1975)
17. J. Lin, R.-J. Lian, Intelligent control of active suspension systems. IEEE Trans. Industr. Electron. **58**(2), 618–628 (2011)
18. J.K. Hedrick, T. Butsuen, Invariant properties of automotive suspensions. Proc. Inst. Mech. Eng, D: J. Autom. Eng. **204**(1), 21–27 (1990)
19. M.W. Sayers, The international road roughness experiment: establishing correlation and a calibration standard for measurements (1986)

Chapter 4
Mechanical Structures: Mathematical Modeling

4.1 Overview

The organization and arrangement of irrelated or distributed elements in a system or an object is defined as mechanical structure. The elements in a mechanical structure can exhibit the characteristics of different parameters. The main parameters which mainly considerable for such structure are mass, elasticity and damping [1]. In order to investigate the feature of a mechanical structure about the mentioned parameters, many factors should be considered and defined. Some of the most important factors which are the basic definition of mechanical structures are described and focused in the following sections. In this chapter, car suspension system has been considered as the main mechanical structure target.

4.1.1 Degrees of Freedom

The number of the parameters which are defining the configuration independently is known as the degree of freedom (DOF). As the other general definition, in order to describe the motion of a system, the coordinate position which is independent is required to be identified. The number of these independent coordinate positions is showing the number of DOF [2].

Vehicle can be considered as a wellknown mechanical system. A vehicle can be defined as a rigid body which has highly stiff suspension and able to move on a two-dimensional space or on a flat plane. The actual degree of freedom for such vehicle is three as it has two components of translation and one angle of rotation. In the case of car suspension system, degree of freedom has been considered as one or two degrees. Many researches around the car suspension as a mechanical structure considered the DOF of suspension system as two. However, in order to performing some complex mathematical models about the mechanical structure, the DOF for suspension system has been assumed as one.

4.1.2 Periodical System Response

When a mechanical system is under excitation by an internal or external forces, it shows some of the properties of a vibratory response. This motion is called periodic motion which might be in regular interval and repetitive or irregular [3]. The mentioned system response is going to be considered in a period of time T. The time for a complete cycle of motion will be considered as T.

4.1.3 Harmonic Motion

Periodic motion in the simplest form of is actually the harmonic motion in which cosine and sine functions as oscillatory functions are utilized to represent the observed or actual motion. The motion described using a continuous *sine* or *cosine* function is referred to as steady state [4]. *w(t)* or actual displacement may be written in the form of following equation [4]:

$$w(t) = |A| \cos(\omega t + \phi) \tag{4.1}$$

here:

$w(t)$ is actual displacement
$|A|$ real amplitude of motion
ω is circular frequency in radians per second
ϕ is an arbitrary phase angle in radians.

(4.1) The superposition is the combination of *sine* and *cosine* functions, and can be shown as follows [5]:

$$w(t) = A_R \cos \omega t - A_I \sin \omega t \tag{4.2}$$

where A_R and A_I are real numbers of amplitude of motion as follows [5]:

$$A_R = |A| \cos \phi, \tag{4.3}$$

$$A_I = |A| \sin \phi, \tag{4.4}$$

The expression of phase angle is the following equation [5]:

$$\phi = \tan^{-1}\left(\frac{A_I}{A_R}\right) \tag{4.5}$$

The constant $|A|$ in (4.1) is related to the constants A_I and A_R in (4.2) by following equation [5]:

$$|A| = \left(A_R^2 + A_I^2\right)^{1/2} \tag{4.6}$$

4.1.4 Frequency

Frequency has been defined as the cycle's amount which is known as hertz, in every second of the motion which represents the period's reciprocal [3]. The frequency can be shown by following equation [6]:

$$f = \frac{1}{T} \tag{4.7}$$

Radians per second which is ω or circular frequency can be shown as following equation [3].

$$\omega = 2\pi f \tag{4.8}$$

4.1.5 Amplitude

The periodical amount of the response (maximum amount) of the system is amplitude. For instance, $|A|$ is the motion's amplitude if the motion is stated by (4.1).

4.1.6 The Mean Square Amplitude

The average of the time of response's square can be defined as the amplitude mean square [3]. So that, for instance, consider the following equation:

$$\overline{w^2} = \lim_{T \to \infty} \frac{1}{T} \int_0^T w^2(t)\mathrm{d}t. \tag{4.9}$$

The positive value of the square root of the mean square amplitude is the root mean square amplitude. In order to have the harmonic oscillation of (4.2), the root mean square amplitude is equal to $A/\sqrt{2}$ and it is independent of the phase.

4.1.7 Free and Forced Vibrations

As the outcome of some initial conditions and without any external disturbances, the free vibrations can be defined as the system's motion [7]. By presence of applying the continues and external disturbance the system's motions will be the defined as Forced vibrations [7].

4.1.8 Phasor

When the system's harmonic motion is represented by a rotating vector is a phasor [3]. The can be represented in the complex formula of the periodic motion of (4.1) and (4.2) which will be more appropriate for manipulating mathematically is shown as follows [8]:

$$w(t) = Ae^{j\omega t}, \qquad (4.10)$$

here $w(t)$ and A are complex with the complex amplitude stated as follows [8]:

$$A = A_R + j A_I \qquad (4.11)$$

Figure 4.1 illustrates the phasor form of (3.8). The real motion's amplitude is $|A|$ and it is the length of the vector. Counter clockwise rotation of the vector with angular velocity, on the imaginary and real axis of ω plan causes projection, changes agreeably with time t changing. A cycle of motion is 360° rotation of the vector.

The actual measured or observed motion is the real phasor's component or the complex description. So, the actual motion is shown in following [9]:

$$w(t) = \text{Re}\left[Ae^{j\omega t}\right]. \qquad (4.12)$$

Fig. 4.1 Harmonic motion's phasor diagram [10]

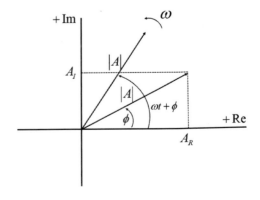

Equation (4.2) yields by utilizing $e^{j\omega t} = \cos \omega t + \sin \omega t$ and substitution of $A = A_R + jA_I$ into the (4.12). As it is shown in (4.5), the real and imaginary ratio of phasor's components will result the ϕ which is the phase of the motion.

The vibration is the main reason of utilizing the negative sign in (4.2). This selection guarantee that the ϕ phase has positive value and meanwhile A_t has a constant value which should be evaluated using boundary conditions making the negative sign choosing to be insignificant against the result.

As the phasor known as vector, same category of harmonic motions with same frequency can be sum up as vector. This is the reason behind complex notation for linear motions, as superposition's principle remains valid, which means separate summation of the imaginary and real components that form motions but individually.

Many responses presented this chapter are written using complex notation because it's a convenient way of analyzing systems having superimposed responses (i.e. like in the case of active controls simulations). The complex descriptions (the real parts) will recover the real motion. In the case that motion has direct description in the actual form, then will be indicated in the subsequent text.

4.2 Single Degree of Freedom Mechanical Systems

As shown in Fig. 4.2, a mass M being supported on a spring with neglectable mass. The w is a unique variable which illustrating the system's displacement, so that the system possesses a single degree of freedom (SDOF) [3, 11]. Followed by appropriate.

By the reason of the gravity, the constant force is neglectable if the resting point of the coordinate system is the origin of the system. The free body diagram of the Fig. 4.2 illustrates that the amount of the w (Positive) which can be achieved by equilibrium position and the mass displacement, force the spring to apply the Kw (restoring force) due to the elongation. Due to the mass releasing, mass acceleration will occur by the spring. The relation between the restoring force and acceleration are shown in the following formula which is based on Newton's second law of motion (see Eq. 4.13).

$$M\frac{d^2w}{dt^2} = -Kw \tag{4.13}$$

The simple single degree of freedom system's motion will be described by differential equation (see Eq. 4.14)

$$\frac{d^2w}{dt^2} + \left(\frac{K}{M}\right)w = 0 \tag{4.14}$$

Equation (4.14) is a second-order ordinary differential equation furthermore, along these lines must have an answer which is indicated as far as two obscure

Fig. 4.2 Undamped and
damped single degree of
freedom systems [11]

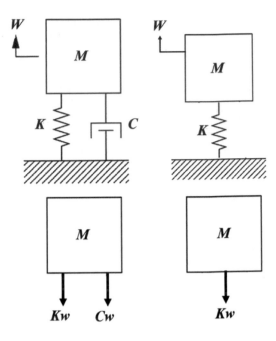

constants or amplitudes of movement. Despite the fact that the above analysis is clear, it illustrates the fundamental procedure in which flexible systems are commonly examined. By dividing the into components (squares). For some underlying conditions the re-establishing and inertial powers are adjusted, subsequently giving the differential condition portraying the movement of the system.

4.3   SODF Free Motion

As the mechanical systems in free motion having harmonic response, (4.14) solution can be form by (4.2). Consequently, the following equation can describe the actual motion [11].

$$w(t) = A_R \cos \omega t - A_I \sin \omega t \tag{4.15}$$

where A_R and A_I are real amplitudes of motion.

In order to determine the condition to have natural vibration, the relation for frequency ω should be provided. By substitution of (4.14) into (4.15) we have can get the following equation [11]:

$$\omega_n = \sqrt{\frac{K}{M}} \tag{4.16}$$

and thus, the solution of (4.14) becomes as follows:

$$w(t) = A_R \cos \omega_n t - A_I \sin \omega_n t \tag{4.17}$$

With considering the (4.17), motion can be completely specified if A_R and A_I being defined as unknown constants. These unknown constants will be defined by utilizing initial or boundary conditions. One of the most essential factors of system's characteristics is ω_n which is resonance or natural frequency. It is noticeable to mentioned that by increasing the K (stiffness) and M (mass) in SDOF systems, the natural frequency will increase. The mentioned states are generally correct if the system is elastic and linear.

To determine the total system's motion, it is mandatory to utilize the introductory conditions first. For instance, if at $t = 0$ the system will have a real velocity w and an underlying genuine removal $\dot{w}$ then the unknown constants in (4.17) can be determined from the following conditions [11]:

$$w(0) = A_R, \tag{4.18}$$

$$\dot{w}(0) = -\omega_n A_I, \tag{4.19}$$

where the use of the overdot is a conservative documentation for separation as for time. The watched reaction will be achieved by fathoming for the A_R and A_I parameters from Eqs. (4.18) and (4.19) and utilizing these two equations into (4.17). The genuine reaction of the system to subjective starting conditions will be obtained by (4.20) to (4.23) [11].

$$w(t) = w(0) \cos \omega_n t + \frac{\dot{w}(0)}{\omega_n} \sin \omega_n t \tag{4.20}$$

It is also possible to write the motion as follows:

$$w(t) = |A| \cos(\omega_n t + \phi) \tag{4.21}$$

in which ϕ (phase angle) is obtained from (4.5) the following:

$$\phi = \tan^{-1} \left(-\frac{\dot{w}(0)}{\omega_n w(0)} \right) \tag{4.22}$$

Equation (4.6) will give and the motion's amplitude, and then the following equation will be the result (see Eq. 4.23)

$$|A| = \left[[w(0)]^2 + \left(\frac{\dot{w}(0)}{\omega_n} \right)^2 \right]^{1/2} \tag{4.23}$$

Thus the response of the single degree of freedom is observable at ω_n (Natural Frequency) as simple harmonic motion, with $|A|$ (amplitude) and ϕ (phase angle) obtained by (4.22) and (4.23) respectively.

4.4   Damped Motion of Single Degree of Freedom Systems

In all of the real systems, the source of vibration is a type of mechanism (damping). Also the vibration energy will be loosen amid the motion cycle. When there is proportional resistant force and the act is in opposite direction of the velocity there is the simples type of damping [3]. Thus, the damping force is specified by the following equation:

$$F_d = -C\frac{dw}{dt},\tag{4.24}$$

where C is the damping coefficient. Figure 4.2 shows a system with a single degree of freedom including the mentioned damping type. This type of damping is known as viscous damping. With considering the additional damping force of new system's force equalization will drive the Eqs. (4.25)–(4.33) [12]. It has been also shown in Fig. 4.2 as the free body diagram.

$$M\frac{d^2w}{dt^2} + C\frac{dw}{dt} + Kw = 0\tag{4.25}$$

It is currently more advantageous to utilize a perplexing depiction of the motion. Accordingly, an answer is accepted of the form the following equation:

$$w(t) = Ae^{\gamma t},\tag{4.26}$$

In this stage $w(t)$ is a variable but complex. γ will be obtained by (4.25) into (4.26) substitution (see Eq. 4.27).

$$\gamma = -\frac{C}{2M} \pm j\sqrt{\frac{K}{M} - \left(\frac{C}{2M}\right)^2}.\tag{4.27}$$

By $C_c = 2M\omega_n$, it is possible to show C in the form of circular damping. By utilizing $\zeta = C/C_c$ the ratio of damping will be achieved. (4.27) then reduces to the following:

$$\gamma = -\omega_n\zeta \pm j\omega_n\sqrt{1 - \zeta^2},\tag{4.28}$$

where, is the natural undamped frequency shown by (4.16) is ω_n.

When $\zeta > 1$, In (4.28) both terms will be real, and it point toward a response which decaying steadily without oscillation. It is the definition of overdamped system. When $\zeta = 1$, it is a system which is named critically damped. Figure 4.3 illustrate that the value of ζ shows the minimum required damping for oscillatory motion prevention. Also this value shows that it is guaranteed in the short time the system will be in rest position again. There will be a real square root and γ will have a negative genuine part if $\zeta < 1$. In addition the root in amplitude is with expanding the time. Therefore, the reaction will waver at a damped characteristic frequency (see 4.29).

$$\omega_d = \omega_n \sqrt{1 - \zeta^2} \tag{4.29}$$

This type of system is known light damping system as auxiliary underdamped system. The watched reaction to the predetermined introductory conditions characterized in Sect. 4.3 is gained by utilizing the (4.25) real part and the initial conditions to solve for the constants A_I and A_R as described before. The actual displacement is shown as the following equation:

$$w(t) = \mathrm{e}^{-\omega_n \zeta t} \left[w(0) \cos \omega_d t + \frac{\dot{w}(0) + \zeta \omega_n w(0)}{\omega_d} \sin \omega_d t \right] \tag{4.30}$$

The simple harmonic form of the equation is as follows:

$$w(t) = |A| \mathrm{e}^{-\omega_n \zeta t} \cos(\omega_d t + \phi), \tag{4.31}$$

ϕ can be written as follows:

$$\phi = \tan^{-1} \left(-\frac{\dot{w}(0) + \zeta w(0)}{w(0)\omega_d} \right) \tag{4.32}$$

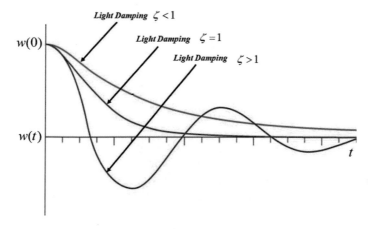

Fig. 4.3 Different value of damping and their response

Then $|A|$ can be written as follows:

$$|A| = \left\{ [w(0)]^2 + [\dot{w}(0) + \zeta \omega_n w(0)]^2 / \omega_d^2 \right\}^{\frac{1}{2}} \tag{4.33}$$

According to (4.31), harmonic motion which is in the frequency of rod and the amplitude as the outcome of the $|A|e^{-\omega_n \zeta t}$ (which is decline by increasing the time) are the constituent of the response. Focusing on the (4.29) showing that the natural frequency is higher than the damped frequency.

In active control models, the damping characteristics and presence are essential. They are also important because they represent a process in which the system response can be decrease by passive means. The existence and feature of the damping is essential for both active and passive control. The SDOF system with different damping and its general response against and time cure has been shown in Fig. 4.3. As the result of the (4.33), a response with oscillating at the rod is occurred by a light damping and its amplitude is decreasing slowly with the time. The response goes near to equilibrium position by a critical damping, but it does not pass it. The respond is going to be constant about the oscillatory motion with heavy damping; the motion will be expressively slow down even near to the equilibrium position by the damping force and it takes time to return to the initial position.

4.5 Forced Response of SDOF Systems

Excitation of most of the systems is with continues disturbance which is going to be applied on the system. This type of excitation is more than the ones mentioned free motion initial excitation. Assume to write the complex form amplitude which is constant and with a harmonic force as the disturbance, shows as follows [13]:

$$f(t) = Fe^{j\omega t} \tag{4.34}$$

Here, the applied force and its relative phase and the amplitude which will be defined by a complex number which is F. (4.14) has to be a homogeneous differential equation so that it should be modified with considering the disturbance force [13].

$$M\frac{d^2 w}{dt^2} + Kw = Fe^{j\omega t} \tag{4.35}$$

In order to state the (4.36), it is noticeable to remember that the disturbance has been assumed to be applied during the whole-time t and in this condition the component of transient response will be zero. So that the response will be considered as steady state and it will be harmonically beginning in a complex form [13].

$$w(t) = Ae^{j\omega t} \tag{4.36}$$

here A and $w(t)$ are complex generally. By substitution of this assumption in (3.33), (4.37) will be obtained as follows [13]:

$$\left(-\omega^2 + \frac{K}{M}\right)A = \frac{F}{M} \tag{4.37}$$

A is an amplitude with complex response and unknown. It can be achieved by reorganizing the (4.37). So that we can obtain the following equation [13]:

$$A = \frac{F/K}{1 - (\omega/\omega_n)^2} \tag{4.38}$$

The response of the elastic systems can be shifted to the disturbance which is harmonic and steady state with the above Eq. (4.38). Theoretically, the displacement amplitude is infinite for the excitations which are steady state and continues, whenever the $\omega = \omega_n$. A large response will be the result of the frequency very close to the natural one during the system driving. From the control point of view, this condition is so essential because the system has been driven on resonance.

The above approach is being so difficult to extension for systems which are complex. For such systems, impedance method is an appropriate method. Here the mechanical input impedance is defined as the complex amplitude ratio of the drive point's force input to velocity. So that, in (4.36) for harmonic displacement, $\dot{w}$ is the velocity such as following [8, 10]:

$$\dot{w}(t) = j\omega A e^{j\omega t} \tag{4.39}$$

The velocity thus has complex amplitude $j\omega A$, and input impedance is given as follows [13]:

$$Z_i = \frac{F}{j\omega A} \tag{4.40}$$

Figure 4.4 illustrates the system input impedance, and it is driven as following:

$$Z_i = \frac{-jK[1 - (\omega/\omega_n)^2]}{\omega} \tag{4.41}$$

By utilizing the (4.40), the system's response to a disturbance force which is harmonic and steady state can be calculated if the input impedance is measured. The amplitude can be measured by (3.42) if we know the relation between the force response and damping (see Eq. (4.42)).

$$A = \frac{F/K}{1 - (\omega/\omega_n)^2 + j2\zeta(\omega/\omega_n)} \tag{4.42}$$

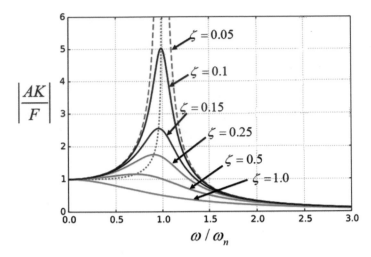

Fig. 4.4 Magnitude of forced response of a single degree of freedom system [13]

Figure 4.5 presents the non-dimensional displacement response amplitude equal to $|AK/F|$ The relationship between the $|AK/F|$ and ω/ω_n which are the non-dimensional displacement response amplitude and input frequency, for different ζ which is the ratio of damping is shown in fig. 4.5. This figure also indicates that the less value of damping will cause large value of response reduction near or on resonance and not that much different than the resonance condition. On the other hand, at resonance, system will be bounded by damping.

Figure 4.5 shows the phase response. The displacement response's phase of the system's excitation force has almost 180° flipping change of the phase for low value of light damping. The reason behind this fact is the incremental behavior of the frequency of the excitation over the resonance frequency. The phase sharpness transaction has the opposite relationship with damping amount. This relationship is important for observation of the system with many degrees of freedom.

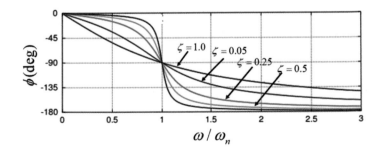

Fig. 4.5 Phases of forced response of a single degree of freedom system [13]

4.6 Car Suspension Types as Mechanical Structures and Performance

As it has been mentioned before, vehicle is considered as the target mechanical structure. On vehicle, the suspension system is focused to be modeled mathematically. By considering the principles of control and control functions, the vehicle suspension system can be categorized into three main categories which are passive, semi-active and active suspensions.

4.6.1 Passive Suspension

While the suspension system has no other power and actuator and the coefficient of damping and stiffness is fixed, the system will be named as passive suspension. Figure 4.6 shows a typical type passive suspension system. As the figure shows, this type of system contains springs, dampers and sprung and unsprung masses. This type of suspension system is considered as traditional mechanical structures which have low cost, reliable performance, no additional energy and simple structure. Passive suspension system is widely utilizing in many types of vehicle. Based on random vibration theory, due to inability of adjusting stiffness and damping in passive suspensions, there is no adoptability for different road and it can only ensure the specific operating conditions to achieve optimal damping effect. Due to this feature, by utilizing this type of suspension good ride comfort and handling stability will be hard to be acquired [14].

Due to the mentioned characteristics of mechanical structures which have been mentioned in the beginning of the chapter, the passive suspension has the following two main defects:

1. It has large travel stroke while the frequency is reducing. The reason behind it was that the natural frequency of the system square is propositional and in reverse direction of the suspension travel stroke.
2. All of the mentioned parameters which are mentioned in the beginning of chapter are restricted due to the suspension components' limitation about the stiffness and damping; in addition, there is no the possibility of meeting with different speed, load, road conditions and so on.

There are a few aspects which might improve the passive suspension systems' performance.

1. Finding the optimal suspension and mechanical structures' parameters by the aid of modeling and simulation;
2. Utilizing the variable-damping shock absorbers and gradient stiffness springs to make the suspension system to be adoptable for different conditions;
3. Utilization of the multi-link suspension including stabilizer bar [16].

Fig. 4.6 Passive suspension
system [15]

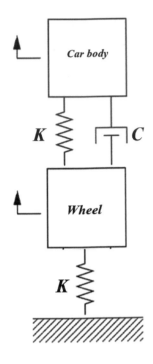

4.6.2   Semi-active Suspension

Whenever there is the possibility of adjusting the shock absorber and elastic element stiffness based on the need, the suspension system will be named as semi-active damping. Semi-active suspension concept was proposed by Karnopp et al. in 1974 with the name of Skyhook semi-active suspension. Due to difficulty of adjusting the stiffness of spring, mainly regulating the shock absorber damping will be focused on semi-active suspensions. Figure 4.7 shows the semi-active suspension systems components.

There is no any specific component of dynamic control in semi-active suspension system. There are some sensors which measuring the velocity and sending the data as signal to car Engine Control Unit (ECU). ECU will measure the input signal and calculate the required control force. In order to compensate the vibration, the signal will be transferred to the shock absorbers to control the damping. Investigation of the actuators (shock absorbers) and the control architecture are two main aspects which have to be focused on semi-active suspensions. In addition, less size of the device and low cost are two advantages of the semi-active suspension in comparison with the active ones [18].

Fig. 4.7 Semi-active
suspension system [17]

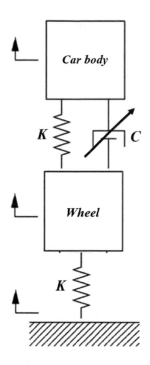

4.6.3 Active Suspension

Due to the need of more accurate and better performance vehicle, different technologies have been utilized to increase the performance of the suspension systems. In this order, different types of manufacturing techniques and technologies, minicomputers, microprocessors, electrohydraulic systems, and different types of controllers have been innovated to be utilized in the suspension systems. To make the vehicle to be smart enough to control itself about the vibration in different road condition and vehicle load, active suspension system will be installed. In order to achieve this goal, an effective control architecture is needed for active suspension system [19]. Figure 4.8 shows an active suspension system.

As the active suspensions are the computer-controlled systems, they have the following features:

(1) Force generation by a power source;
(2) Utilized components to be able to be functioning continuously and passing the mentioned force;
(3) Utilized different type of sensors and huge amount of data sets.

With considering the mentioned features, this type of suspension systems required mechanics, electronics, and control knowledge together. In braking and turning conditions, the springs will be deformed and the inertia force will be generated. The

Fig. 4.8 ¼ model of active
suspension system [20]

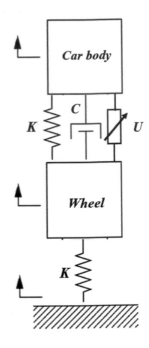

active suspension system will generate a different type of inertia force to compensate the one generated due to the braking and turning to reduce the care position changing [21].

In Table 4.1, a comparison of different type of car suspension system is illustrated.

Table 4.1 Three types of suspension systems' technical comparison [22]

Suspension system type	Passive	Semi-active	Active
Regulatory element	General shock absorber	Adjustable damper	Hydraulic system or servo motor system
Action principle	Damping constant	Damping continuously adjust	Adjust the force between wheel and vehicle body
Control method	No	Electronic-hydraulic automatic	Electronics or magnetic or fluid control
Bandwidth	Unknown	Up to 20 Hz	>15 Hz
Energy consumption	Zero	Very small	Big
Lateral dynamics	No	Middle	Good
Vertical dynamics	No	Middle	Good
Cost	Lowest	Middle	Highest

In this chapter, it has been focused to just model the active and passive suspension systems. No matter how and what kind, only the amount of created force is having been discussed.

4.6.4   Mathematical Modeling of Passive Suspension System

Mathematical modeling of the system and determining the requirements for its design are the first step in every system design. In order to evaluate the designed control system, several models will be tested by simulation techniques. This procedure shows the importance of the mathematical modeling of the system which is as the prerequisite for the control architecture and its design [23–39].

Providing the model description by parameters determination is the key to create a mathematical model of a system.

There are three types of well-known models in the case of vibration control (here suspension control) which are weight function, the state space, and transfer function description. It is possible to categorize them as the continues and discrete time mathematical models. In order to model the passive suspension system mathematically, it is considered to start the procedure with simplifying the system as it has been illustrated in Fig. 4.6. In addition, adding more detail for it to be capable of being utilized in mathematical model (Fig. 4.9).

In Fig. 4.9 which illustrating a passive suspension in a quarter vehicle model, there are sprung mass M_1 and unsprung mass M_2 which are vehicle body and an assembly

Fig. 4.9   Vehicle ¼ passive suspension system mathematical model [15]

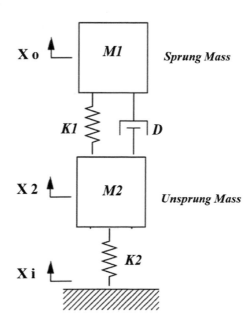

of the wheel and axle, respectively. The linear spring with stiffness k_2 represents the tire when is contacting the road during the car is traveling. The main components of the passive suspension system are damper and spring. D represent the linear damper with average coefficient of damping D and a linear actual spring with coefficient of stiffness k_1. $x_0(t)$ shows the displacement of the vehicle body (sprung mass), and $x_2(t)$ is representing the displacement of wheel and axle (unsprung Mass). $x_i(t)$ is representing the vertical road profile.

Unlike the definition and mathematical modeling which are explained in the beginning of this chapter (SDOF), here there is a different mathematical modeling as this system has two degrees of freedom. So that the vehicle suspension system's dynamics and two differential equations of degrees of freedom motion will be represented as follows [40]:

$$M_1\ddot{x}_0(t) + D[\dot{x}_0(t) - \dot{x}_2(t)] + k_1[x_0(t) - x_2(t)] = 0 \tag{4.43}$$

$$M_2\ddot{x}_2(t) - D[\dot{x}_0(t) - \dot{x}_2(t)] + k_1[x_2(t) - x_0(t)] + k_2[x_2(t) - x_i(t)] = 0 \tag{4.44}$$

By applying the Laplace transformation on (4.44), we can get (4.45) and it is passive suspension system transfer function [40].

$$\frac{X_0}{X_1} = \frac{k_2(Ds + k_1)}{M_1M_2s^4 + (M_1 + M_2)Ds^3 + (M_1k_1 + M_1k_2 + M_2k_1)s^2 + Dk_2s + k_1k_2} \tag{4.45}$$

In order to do simulation by utilizing the mentioned mathematical model, the value of parameters which could be different for each car should be defined. Here is an example of these parameters' definition which has been given as Table 4.2.

Table 4.2 Vehicle ¼ passive suspension model simulation input parameters [22, 41]

Vehicle model parameters	Symbol	Numerical value	Unit
Sprung mass	M_1	300	kg
Unsprung mass	M_2	40	kg
Suspension stiffness	K_1	15,000	N/m
Tire stiffness	K_2	150,000	N/m
Suspension damping coefficient	D	1000	N^*s/m

4.6.5 *Mathematical Modeling Active Suspension System*

In this chapter, it has been focused to just model the active and passive suspension systems. No matter how and what kind, only the amount of created force is having been discussed.

In order to utilize pavement method and its roughness as input for modeling of the body of vehicle in vertical vibration, ¼ vehicle dynamic vibration model has been chosen. This model dose not include all of the features such as geometrical information and pitching and rolling angle vibration, but it is still covering almost all of the features such as the load and suspension system's stress change information. Here, the characteristics of the ¼ vehicle dynamic vibration model have been provided [42]:

(1) Two sets of suspension system installed in the vehicle which are the suspensions for front side and back side are independently functioning from each other of each other.
(2) The suspension itself and the tires have to be investigated together.
(3) The components which are elastic can be simply can be assumed as damping and springs.
(4) Decrease the system parametric description under precondition of keeping accuracy and effectiveness.
(5) Mass of the actuator will not be considered, and force is their only output.

Figure 4.10 illustrates a vehicle ¼ model of active suspension system. Everything about the parameters in this figure is same as passive suspension's assumptions. The only different is the u which is the active control force and has been generated by the actuator of active suspension.

Form Fig. 4.10, dynamics of the vehicle suspension system can be analyzed. Also, it is possible to create two differential equations of degrees of freedom motion as follows [23]:

$$M_1\ddot{x}_0(t) + D[\dot{x}_0(t) - \dot{x}_2(t)] + k_1[x_0(t) - x_2(t)] = \text{u} \qquad (4.46)$$

$$M_2\ddot{x}_2(t) - D[\dot{x}_0(t) - \dot{x}_2(t)] + k_1[x_2(t) - x_0(t)] + k_2[x_2(t) - x_1(t)] = -u \quad (4.47)$$

Equations. (4.48) to (4.58) illustrate the mathematical modeling of an active suspension system with state space method [22, 23].

$$x_1 = x_2(t), x_2 = x_0(t), x_3 = \dot{x}_2(t), x_4 = \dot{x}_0(t) \qquad (4.48)$$

Equation of the system state space can be shown as follows:

$$\frac{dX}{dt} = AX + BU \qquad (4.49)$$

In (4.49), state variable matrixes can be shown as follows:

Fig. 4.10 ¼ model of active
suspension system [20]

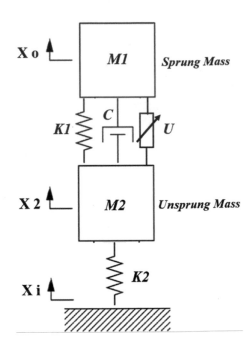

$$x = [\, x_1 \; x_2 \; x_3 \; x_4 \,] \qquad (4.50)$$

A and B are constant matrixes, and they can be shown as follows:

$$\begin{bmatrix} 0 & 0 & 1 & 0 \\ 0 & 0 & 0 & 1 \\ \frac{K_1+K_2}{m_2} & \frac{K_1}{m_2} & \frac{D}{m_2} & \frac{D}{m_2} \\ \frac{K_1}{m_1} & \frac{K_1}{m_1} & \frac{D}{m_1} & \frac{D}{m_1} \end{bmatrix} \qquad (4.51)$$

$$\begin{bmatrix} 0 & 0 \\ 0 & 0 \\ \frac{K_2}{m_2} & \frac{1}{m_2} \\ 0 & -\frac{1}{m_1} \end{bmatrix} \qquad (4.52)$$

The matrix belongs to the input variable of the system which is as follows:

$$U = [\, x_i(t) \; u \,]^{\mathrm{T}} \qquad (4.53)$$

Output matrix of the suspension system is shown as follows:

$$Y = CX + DU \qquad (4.54)$$

In (4.45), Y is the output variable matrix and it is shown as follows:

$$Y = \{k_2[x_i(t) - x_1] \quad \ddot{x}_0(t) \quad x_0(t)\} \tag{4.55}$$

The following equation is another representation of Y:

$$Y = \{k_2[x_i(t) - x_1] \quad \ddot{x}_2 \quad x_2\} \tag{4.56}$$

C and D are constant matrixes, and they can be shown as follows:

$$\begin{bmatrix} -k_2 & 0 & 0 & 0 \\ \frac{k_1}{m_1} & -\frac{k_1}{m_1} & \frac{D}{m_1} & \frac{D}{m_1} \\ 0 & 1 & 0 & 0 \end{bmatrix} \tag{4.57}$$

$$\begin{bmatrix} k_2 & 0 \\ 0 & -\frac{1}{m_1} \\ 0 & 0 \end{bmatrix} \tag{4.58}$$

Eliminating the pavement condition by utilizing a controller was the main goal of this section. It means adjusting the force which is generated by active suspension system to compensate the road condition changes. It is worthy to mention that the road excitation and changes can be assumed as noise based on the definitions in previous chapters. To achieve the goal of this chapter, in the next chapter definition of noise cancelation and its common methods will be focused.

References

1. S.R. Singiresu, *Mechanical Vibrations* (Addison Wesley, 1995)
2. J.G. Eisenhauer, Degrees of freedom. Teaching Statistics **30**(3), 75–78 (2008)
3. R.F. Steidel, *An introduction to mechanical vibrations* (Wiley, New York, 1979)
4. G.R. Fowles, G.L. Cassiday, *Analytical Mechanics* (Saunders College, 1999)
5. A.P. French, *Vibrations and Waves* (CRC press, 1971)
6. W. Matthaeus, M. Goldstein, Low-frequency 1 f noise in the interplanetary magnetic field. Phys. Rev. Lett. **57**(4), 495 (1986)
7. S. Sarkani, L.D. Lutes, *Stochastic Analysis of Structural and Mechanical Vibrations* (Prentice Hall, 1997)
8. M. Feldman, Hilbert transform in vibration analysis. Mech. Syst. Signal Process. **25**(3), 735–802 (2011)
9. A.G. Phadke, J. S. Thorp, *Synchronized Phasor Measurements and Their Applications* (Springer, 2008)
10. C.C. Fuller, S. Elliott, P.A. Nelson, *Active Control of Vibration* (Academic Press, 1996)
11. A.K. Chopra, *Dynamics of Structures. Theory and Applications to Earthquake Engineering* (2017)
12. K. Ogata, *System Dynamics* (Prentice Hall, Upper Saddle River, NJ, 1998)
13. S.C. Arya, M.W. O'neill, G. Pincus, *Design of Structures and Foundations for Vibrating Machines* (Gulf Publishing Company, Books Division, 1979)

14. M.M. Fateh, S.S. Alavi, Impedance control of an active suspension system. Mechatronics **19**(1), 134–140 (2009)
15. J. Tamboli, S. Joshi, Optimum design of a passive suspension system of a vehicle subjected to actual random road excitations. J. Sound Vib. **219**(2), 193–205 (1999)
16. C.L. Phillips, H.T. Nagle, *Digital Control System Analysis and Design* (Prentice Hall Press, 2007)
17. A. Ahmad, Y.M. Sam, N.M.A. Ghani, F.K. Elektrik, *An Observer Design for Active Suspension System* (Universiti Teknologi Malaysia, 2005)
18. X. Xue et al., in *Study of Art of Automotive Active Suspensions*. Power Electronics Systems and Applications (PESA), 2011 4th International Conference on (IEEE, 2011), pp. 1–7
19. N. Yagiz, Y. Hacioglu, Backstepping control of a vehicle with active suspensions. Control Eng. Pract. **16**(12), 1457–1467 (2008)
20. A. Agharkakli, G.S. Sabet, A. Barouz, Simulation and analysis of passive and active suspension system using quarter car model for different road profile. Int. J. Eng. Trends Technol. **3**(5), 636–644 (2012)
21. A. Gupta, J. Jendrzejczyk, T. Mulcahy, J. Hull, Design of electromagnetic shock absorbers. Int. J. Mech. Mater. Des. **3**(3), 285–291 (2006)
22. Q. Zhou, *Research and Simulation on New Active Suspension Control System* (2013)
23. A. Azizi, Computer-based analysis of the stochastic stability of mechanical structures driven by white and colored noise. Sustainability **10**(10), 3419 (2018)
24. A. Ashkzari, A. Azizi, Introducing genetic algorithm as an intelligent optimization technique, in *Applied Mechanics and Materials*, vol. 568 (Trans Tech Publications, 2014), pp. 793–797
25. A. Azizi, Introducing a novel hybrid artificial intelligence algorithm to optimize network of industrial applications in modern manufacturing. Complexity 2017 (2017)
26. A. Azizi, Hybrid artificial intelligence optimization technique, in *Applications of Artificial Intelligence Techniques in Industry 4.0* (Springer, 2019), pp. 27–47
27. A. Azizi, Modern Manufacturing, in *Applications of Artificial Intelligence Techniques in Industry 4.0* (Springer, 2019), pp. 7–17
28. A. Azizi, RFID Network Planning, in *Applications of Artificial Intelligence Techniques in Industry 4.0* (Springer, 2019), pp. 19–25
29. A. Azizi, *Applications of Artificial Intelligence Techniques in Industry 4.0* (Springer)
30. A. Azizi, F. Entesari, K.G. Osgouie, M. Cheragh, Intelligent Mobile Robot Navigation in an Uncertain Dynamic Environment, in *Applied Mechanics and Materials*, vol. 367(Trans Tech Publications, 2013), pp. 388–392
31. A. Azizi, F. Entessari, K.G. Osgouie, A.R. Rashnoodi, Introducing neural networks as a computational intelligent technique, in *Applied Mechanics and Materials*, vol. 464 (Trans Tech Publications, 2014), pp. 369–374
32. A. Azizi, N. Seifipour, *Modeling of Dermal wound Healing-Remodeling Phase by Neural Networks*, in Computer Science and Information Technology-Spring Conference, 2009. IAC-SITSC'09. International Association of (IEEE, 2009), pp. 447–450
33. A. Azizi, A. Vatankhah Barenji, M. Hashmipour, Optimizing radio frequency identification network planning through ring probabilistic logic neurons. Adv. Mech. Eng. **8**(8), 1687814016663476 (2016)
34. A. Azizi, P.G. Yazdi, M. Hashemipour, Interactive design of storage unit utilizing virtual reality and ergonomic framework for production optimization in manufacturing industry. Int. J. Interact. Des. Manuf. (IJIDeM), 1–9 (2018)
35. M. Koopialipoor, A. Fallah, D.J. Armaghani, A. Azizi, E.T. Mohamad, Three hybrid intelligent models in estimating flyrock distance resulting from blasting. Eng. Comput., 1–14 (2018)
36. K.G. Osgouie, A. Azizi, in *Optimizing Fuzzy Logic Controller for Diabetes Type I by Genetic Algorithm*. Computer and Automation Engineering (ICCAE), 2010 The 2nd International Conference on, vol. 2 (IEEE, 2010), pp. 4–8
37. S. Rashidnejhad, A.H. Asfia, K.G. Osgouie, A. Meghdari, A. Azizi, Optimal trajectory planning for parallel robots considering time-jerk, in *Applied Mechanics and Materials*, vol. 390 (Trans Tech Publications, 2013), pp. 471–477

38. J. Wang, W. Wang, K. Atallah, D. Howe, in *Design of a Linear Permanent Magnet Motor for Active Vehicle Suspension*. Electric Machines and Drives Conference, 2009. IEMDC'09. IEEE International (IEEE, 2009), pp. 585–591
39. J. Wang, W. Wang, K. Atallah, A linear permanent-magnet motor for active vehicle suspension. IEEE Trans. Veh. Technol. **60**(1), 55–63 (2011)
40. M.S. Kumar, *Development of active Suspension System for Automobiles Using PID Controller* (2008)
41. M. Zhou, H. Jin, W. Wang, A review of vehicle fuel consumption models to evaluate eco-driving and eco-routing. Transp. Res. D: Transp. Environ. **49**, 203–218 (2016)
42. M.A. Nekoui, P. Hadavi, in *Optimal Control of an Active Suspension System*. Power Electronics and Motion Control Conference (EPE/PEMC), 2010 14th International (IEEE, 2010), pp. T5-60–T5-63

Chapter 5
Noise Control Techniques

5.1 Overview

In the scientific terminology, noise control is an operation which involves filtering, canceling or reducing out the unwanted noise or interference from the signal. So that, by the aid of noise control, desired signal can be recovered. Noise reduction increases the signal-to-noise ratio, and it is directly affecting the different parameters of the structure model [1, 2].

There are several controllers that have been proposed in recent years to control the linear and nonlinear systems [3–17]. In this chapter, it has been tried to focus on three well-known controllers and control method which are adaptive noise control including adaptive noise filtration and cancelation [18], sliding mode [19], and PID control [20].

5.2 Adaptive Noise Filter

In order to modeling the real-time relationship of two signals continuously, a computational device is needed which is named as adaptive filter. Usually, adoptive filters are known as a category of instructions of programs which are functioning on a processing unit or a category of logical operations executed in field-programmable gate array (FPGA). The error ignorance about the precision (numerical) in these categories will affect the execution of them. There is an independency between physical realization of the adoptive filters and its fundamental operations. Due to the mentioned relationship, adoptive filters are going to be focused mathematically rather than their hardware and software realizations. There are main aspects for adoptive filter [21]:

© The Author(s), under exclusive license to Springer Nature Singapore Pte Ltd. 2019   61
A. Azizi and P. G. Yazdi, *Computer-Based Analysis of the Stochastic Stability of Mechanical Structures Driven by White and Colored Noise*, SpringerBriefs in Applied Sciences and Technology, https://doi.org/10.1007/978-981-13-6218-7_5

Fig. 5.1 General block
diagram of adaptive filter
[22]

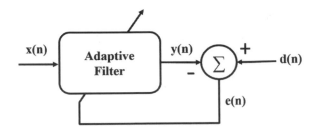

- A filter is processing the signals.
- The structure of the filter indicated that how the input signal is computed.
- The filter structure's parameters can continuously change to change the relation between input and output.
- The algorithm structure shows the way of parameters adjustments.

With considering the Fig. 5.1, it is obvious that the devices shown as the adoptive filter have input signals (digital) named $x(n)$ and compute the output signal (digital) named $y(n)$ in the period of time n. For the first stage, the only important item about the adoptive filter structure is their parameters which will affect the output $y(n)$ computation. By assuming the desire response signal $d(n)$ as a second signal and output signal $y(n)$, it is possible to evaluate the difference signal. This signal can be written as (5.1), and it named as error signal [see Eq. (5.1)].

$$e(n) = d(n) - y(n) \tag{5.1}$$

The reason behind the use of the error signal is altering or adapting of the filter parameters between time n to time $n + 1$. The error signal is shown in Fig. 5.1 as the arrow (oblique). It is expected to have decreased manner of the error signal because the main aim of the adoptive filter is to improve the output filter during the time increments to reach to the desire response. This improvement of the output signal and decreasing the error signal will be achieved by adjustment of the filter parameters and modification of the algorithm of the filter.

By changing the parameter of the filter between $t = n$ and $t = n + 1$, a method will be generated to accomplish the adoptive filter task. The selected computational structure for the system will affect the type and number of the require parameters in the filter structure.

Figure 5.1 shows that the number of the parameters which will effect on the computation of $y(n)$ should be a finite number. So that any system with this finite number of the parameter could be an adoptive filter. $W(n)$ is a vector, and it is defined as follows [23]:

$$W(n) = [w_0(n)w_1(n)\ldots w_{L-1}(n)]^T \tag{5.2}$$

Here $\{w_i(n)\}, 0 \leq i \leq L - 1$; there are L system's parameters in time n. The relationship between the input signal and output signal can be defined as follows [24]:

$$y(n) = f(W(n), y(n - 2), \ldots, y(n - N), x(n), x(n - 1), \ldots, x(n - M + 1)) \tag{5.3}$$

Here M and N are integer numbers (positive), and $f(.)$ can be any nonlinear or linear function represents any well-defined. According to (5.3), the filter is casual whenever to compute $y(n)$; the future values of $x(n)$ are not needed.

Equation (5.3) represents a general structure of the adoptive filter but it has been aimed to find the best general linear relationship between the input signals and desired response for variety of problems. Infinite impulse response (IIR) or finite impulse response (FIR) filters should be utilized to achieve the mentioned relationship.

The general structure of the FIR filter is shown in Fig. 5.2 which could be named as transversal and tapped delay line filters.

$z - 1$ represents the unit delay element.

$w_i(n)$ represents the system multiplicative gain.

$W(n)$ is the impulse response in time (n).

The equation can be rewritten about the $y(n)$ as follows [25]:

$$y(n) = \sum_{i=0}^{L-1} w_i(n)x(n - i) \tag{5.4}$$

$$W^T(n)X(n) \tag{5.5}$$

Here $X(n) = [x(n)x(n-1)\ldots x(n-L+1)]^T$ represents the vector of input signal, and $.^T$ represents vector transpose. The mentioned system needs L multiplies and $L - 1$ adds to be implemented. As long as the period for the signals in large and L is short by a processor or a circuit, preforming the computations is easy. In order to

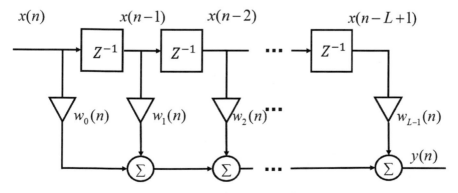

Fig. 5.2 FIR filter structure [26]

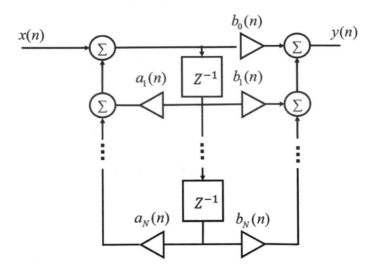

Fig. 5.3 Structure of the infinite impulse response filter [27]

perform the computation, totally 2 L of memory is required, one of them to store L inputs and the other one values of the L coefficient.

Figure 5.2 shows the direct form of the finite impulse response structure. Mathematical representation of the output signal is as follows [25]:

$$y(n) = \sum_{i=1}^{N} a_i(n)y(n-i) + \sum_{j=0}^{N} b_j(n)x(n-j) \qquad (5.6)$$

Figure 5.2 cannot represent the system with considering the mentioned fashion. However, (5.6) can be rewritten by utilizing vector notation [23].

$$y(n) = W^{\mathrm{T}}(n)U(n), \qquad (5.7)$$

In here $U(n)$ and $W(n)$ which are the $(2N + 1)$-dimensional vectors can be represent as follows [23]:

$$W(n) = [a_1(n)a_2(n)\ldots a_N(n)b_0(n)b_1(n)\ldots b_N(n)]^{\mathrm{T}} \qquad (5.8)$$

$$U(n) = [y(n-1)y(n-2)\ldots y(n-N)x(n)x(n-1)\ldots x(n-N)]^{\mathrm{T}}, \qquad (5.9)$$

Unlike the structure of the FIR filter, the number of the adds, multiples, and memory locations for computing the $y(n)$ in IIR filter structures have to be fixed (Fig. 5.3).

Lattice filters are another proper category to do the task of adoptive filters. These types of filter are actually a category of FIR which $L - 1$ processing stage will be utilized to do the computation for a category of auxiliary signals. These signals can be represented as $\{b_i(n)\}$, $0 \leq i \leq L - 1$ which named as backward prediction errors. $X(n)$ elements can be represented through a linear transformation by auxiliary signals as they are uncorrelated. This type of filter can be used when there is delayed input signal and, in the structures, similar to the Fig. 5.2. Due to the uncorrelated property of this type of filters, they can provide better algorithm and improve the filter convergence performance [28].

One of the items which has to be considered in selection of the adoptive filters is the level of complex computation. Due to the real-time operation of the adoptive filters, all of the calculation should be done for one-time sample for the system. All of the mentioned structured are going to be useful since it is possible to compute $y(n)$ in a finite time and memory.

When the superposition's principle does not hold and the value of the parameters is fixed, a nonlinear structure should be considered instead of a linear one. The mentioned type of system is appropriate, whenever the relationship between the $x(n)$ and $d(n)$ has nonlinear feature. Bilinear and Volterra are such filters which can compute the output $y(n)$ according to the past outputs and polynomial representation of the input [29]. In the field of neural network, most of the nonlinear models can fit (5.5) and many of the algorithms for parameter adjustments in this field are related to IIR and FIR adoptive filters.

5.3 Adoptive Noise Cancelation

Active Noise reduction (ANR) or Active Noise Control (ANC) are also known as Noise cancelation. Actually, it is a system which can cancel the Primary Noise (Unwanted Noise) preforming on the superposition principles [30]. Active noise control utilizes proper secondary sources generating canceling or anti-noise waves. There will be a system (electronic) with a specific algorithm to process the signal interconnected to the secondary resources. There are a wide range of available applications of ANC such as industrial and manufacturing operations and so on [19, 31].

Noise filters which are mentioned before can be fixed or adaptive. In order to design the fixed filters, it essential to know and have the knowledge of noise and signal together. On the other hand, due to the ability of the self-adjustment of the parameters in adaptive filters, there is less or no need to have the knowledge of noise or signal in designing of these types of filter. The reality of canceling a noise is actually the optimal filtering and its variation. There will be one or more than one sensor in the noise field where there is weak or not detectable signal in these points. Noise cancelation makes the use of an input which is reference or auxiliary generated by the sensors. So that by subtracting and filtering this input from the primary one which has both noise and signal, the primary noise will be canceled or eliminated. It seems to be so risky procedure to subtract the noise from the signal and any mistake will increase the noise power in the output signal. Although by utilizing a proper adoptive process, subtraction and filtering of the noise will be controlled, but still it has the risk of signal distorting and noise increments in output signal. In some cases, when it is possible to utilize the adaptive noise cancelation, the level of noise rejection will be increased. This makes it nearly impossible to reach to that by direct filters. Here, the adaptive noise cancelation concept with its advantageous and limitations are described [32].

As Fig. 5.4 shows through a channel, a signal (s) is spread to sensor. This channel also receives the noise which is not correlated with the signal. $s + x_0$ shows the combination of the signal and noise which is from the primary input to the noise canceler. Another sensor is getting the noise x_1. This noise is uncorrelated with the input signal but in the same time with the noise x_0. The main function of this sensor is preparing the reference point for the noise canceler. The input which is reference to the canceler will be provided by this sensor. The $y(n)$ which is the output will be produced by filtering x_1, and it is very similar to x_0. The output $y(n)$ will be deducted from the $s + x_0$ which is primary input of the canceler, and the result will be $z(n)$. So that, the $z(n) = s + x_0 - y(n)$. In order to converting x_1 to x_0, it is required to design a filter which is fixed. To design this fixed filter, it is required to know the characteristic of the channel for transferring the noise to the referenced sensor signal. $y(n)$ which is the output of the filter can be deducted from the primary input and the signal will be equal to $z(n)$ which is the output of the system. Fixed filter utilization is no feasible because of the unknown characteristic of the transmission channel. The channel usually does not have fixed nature. On the other hand, even if there is the feasibility chance of utilizing the fixed filter, many other difficulties will be shown up. The precise adjustment of its characteristic is the major difficulty, and it causes error by increasing the noise power in the system output. Figure 5.4 shows that instead of utilizing a fixed filter by considering its utilization difficulties, adoptive filter can process the reference input. This advantage comes from the auto-adjustment ability of the impulse response. There is a specific algorithm to respond to the error signal going to the filter from the system output. With this algorithm, in different conditions filter can check itself to decrease the error signal to minimum [32].

By changing the application, the error signal may vary. Practically, producing the system output ($z(n)$) which is the most fitted one with signal s in the least square sense is the main objective of noise canceling. In order to accomplish the objective of the noise canceling, the system output should go back to the adoptive filter. Least mean square (LMS) should might be the best option as the integrated algorithm inside the adoptive filter to minimize total output power of the system. Actually, in this type of noise canceling system output will be considered as the error signal feed to the adoptive process. Having enough knowledge of the input signal (s) and noises (x_0 and x_1) is essential before the filter adoption and design. Equations (5.10)–(5.15) [32, 34–36] describe the importance of having the knowledge of the system parameters (s, x_0, or x_1), and their relationships, deterministically or statistically [32].

Considering that $y(n)$, s, x_1, and x_0 have zero means and they are stationary. In addition, by considering the correlation between x_1 and x_0 and they are uncorrelated with s, then the output $z(n)$ is as follows [34]:

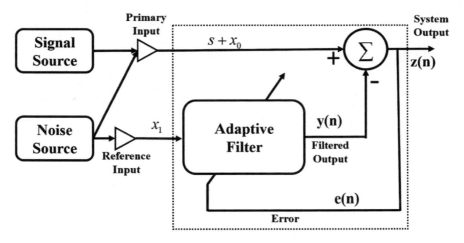

Fig. 5.4 Concept of ANC [33]

$$z = s + x_0 - y(n) \tag{5.10}$$

By squaring (5.10), we can obtain the following [35]:

$$z^2 = s^2 + (x_0 - y)^2 + 2s(x_0 - y) \tag{5.11}$$

Both sides of Eq. (5.11) will get expectations. With considering the uncorrelation relationship between s and x_0 and y, the following equation will be obtained [34]:

$$E[z^2] = E[s^2] + E[(x_0 - y)^2] + 2E[s(x_0 - y)] \tag{5.12}$$

$$= E[s^2] + E[(x_0 - y)^2] \tag{5.13}$$

Both sides of the Eq. (5.12) will be considered as minimum. Whenever the filter will be adjusted to minimize the value of the output power, the signal power $E[s^2]$ will not be affected. So the following equation will be obtained [35]:

$$\min E[z^2] = E[s^2] + \min E[(x_0 - y)^2] \tag{5.14}$$

By minimizing the $E[z^2]$ with filter adjustment, $E[(x_0 - y)^2]$ also will be minimized. In this case, x_0 (primary noise) least square estimate will be with the best value. With considering the Eq. (5.10), by minimizing $E[(x_0 - y)^2]$, $E[(x_0 - y)^2]$ will be minimized. So that, the following equation will be obtained [35, 36]:

$$(z - s) = (x_0 - y) \tag{5.15}$$

As it is mentioned before, filter adjustment to minimizing the output power causes output z to be the best least square estimate of the signal s. Noise and signal s are the constituent of the output z. (5.14) shows that whenever $E[z^2] = E[s^2]$, we have the smallest output power. It is also showing we can achieve the smallest output power when $E[(x_0 - y)^2] = 0$. So that, $y = x_0$, and also $z = s$. With considering the mentioned condition, it is possible to have perfect noise-free output signal by minimizing the output power. The mentioned arguments can be extended to the case when x_0 and x_1 are stochastic and deterministic.

5.4   PID Controller

One of the well-known controllers which has been adopted in many industries is the proportional–integral–derivative (PID) controller. PID controller is a closed-loop controller type which controls the plant output variable by minimizing the error between real plant output and desired output. PID controller consists of three controller modes: P as proportional controller, I as integral controller, and D as derivative controller (see Fig. 5.5) [37]. Before introducing the mathematical model of the PID controller, each component of the controller and combination of them has been discussed in next sections.

5.4.1   P Controller

In industrial control systems, the main mode of the PID controller mostly is known as proportional control mode which determines the controller response to the plant error by multiplying the error to the P controller's gain (K_p), so it means that the higher K_p will result in higher P action to the plant error (5.16). [38]

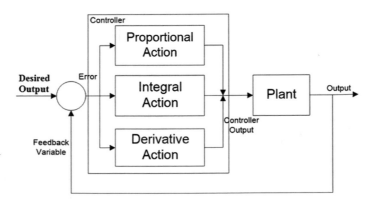

Fig. 5.5   PID controller [3]

$$P = K_p \times e(t) \tag{5.16}$$

One of the important applications of utilizing the P controller is decreasing the steady state error of the systems to stabilize an unstable process of first-order systems which have single energy storage. It is important to know that just by utilizing proportional controllers the steady state error of the system cannot be eliminated. By increasing the K_p, the system response will have smaller amplitude and phase margin, faster dynamics satisfying, wider frequency band, and larger sensitivity to the noise. In addition, it can be concluded that utilizing proportional controller decreases the rise time and after a certain value of reduction on the steady state error, and increasing the gain only leads to overshoot of the system response. Also, it has been observed that they can be the cause of fluctuations in the presence of lags, and it means that process noise will be amplified in higher-order systems.

5.4.2 I Controller

The effect of the integral controller mode can be defined as decreasing or increasing the response time of the controller to the plant error by calculating the integral of the error and multiplying it to the I controller's gain (K_I), so it means that higher K_I results in higher I action to the plant error (5.17) [39].

$$I = K_I \times \int e(t) dt \tag{5.17}$$

5.4.3 D Controller

The last mode of a PID controller which regulates the plant's output by calculating the derivative of the error and multiplying it to the P controller's gain (K_D) is derivative controller mode (5.18). The D mode controllers are widely used in motion control systems since they are very sensitive against noise and disturbances [40].

$$D = K_D \times \frac{de(t)}{dt} \tag{5.18}$$

5.4.4 PI Controller

Steady state error which has been produced by proportional controller can be eliminated by adding integral controller to the proportional one which is known as PI controller. It is important to know that such controllers have a negative impact on the system from the point view of the speed of the response and overall stability of the system. The fluctuations cannot be eliminated by this controller type; also it is not able to increase the rise time and any amount of I guarantees set point overshoot.

Table 5.1 Response of PID controller [3]

Parameter	Stability	Steady state error	Settling time	Overshoot	Rise time
$\uparrow K_P$	Degrade	Decrease	Small change	Increase	Decrease
$\uparrow K_I$	Degrade	Eliminate	Increase	Increase	Decrease
$\uparrow K_D$	Improve for small K_D	No effect in theory	Decrease	Decrease	Minor change

5.4.5 PD Control

The most important property of such a controller is the ability of predicting the system response error, so it can be utilized as a tool to increase the stability of the system.

It is important to mention that since the derivative is taken from the output response of the system, D mode is designed to be proportional to the change of the output variable.

5.4.6 Mathematical Modeling of the PID Controller

It is important to know that PID controller is the weighted sum of the three P, I, and D control modes and based on the plant requirements one or two mode can be eliminated. The response of the PID controller, control signal $u(t)$, to the plant error can be determined as shown in (5.19) [41]:

$$u(t) = \left(K_p \times e(t) \right) + (K_I \times \int e(t)dt) + \left(K_D \times \frac{de(t)}{dt} \right) \qquad (5.19)$$

It is highly important to mention that knowing effect of the each of these three modes on the response of the controller is an essential criterion in control theory; any change in PID controller's coefficients can result in changing the status of the system from stable to unstable (Table 5.1) [41].

As seen in Table 5.1, if the K_p is increased too much, the control loop will begin oscillating and become unstable; also the system will not receive desired control response if K_p is set too low. The similar rules exist for integral and derivative controller modes:

The controller response will be very slow if the integral time is set too long period. In addition, the system will be unstable and the control loop will oscillate if K_I is set too low. However, if K_D will be increased too much then oscillations will occur, and the control loop will turn to unstable as well.

5.5 Sliding Mode Control

The theory of variable structure control (VSC) with sliding mode control (SMC) was proposed for the first time in the early 1950s. Generally, sliding mode controller can be defined as a type of the variable structure controller which has robustness against uncertainties. The sliding word refers to the defined sliding surfaces. In this type of the controller, system behavior should be restricted on the defined surfaces [42].

To design a sliding mode controller, it is necessary to perform two actions: designing a stable sliding surface and designing an effective control law. Such a controller executes fast control actions which make the system stay on designed sliding surface [43].

5.5.1 Mathematical Model of the Sliding Mode Controller

Sliding mode variable structure system consists of a set of continuous subsystems, control actions, and reference inputs. In general sliding modes, nonlinear affine systems can be defined as follows [44]:

$$\dot{x} = f(x, t) + B(x, t)u \tag{5.20}$$

$$u_i = \begin{cases} u_i^+(x, t), s_i(x) > 0 \\ u_i^-(x, t), s_i(x) < 0 \end{cases} \quad i = 1, \ldots, m \tag{5.21}$$

where $x \in R^n$ is a state vector, $u \in R^m$ is a control vector, $u_i^+(x, t), u_i^-(x, t)$, and $s_i(x)$ are continuous functions of their arguments, $u_i^+(x, t) \neq u_i^-(x, t)$. It is important to know that the control is designed as a discontinuous function of the state such that each component undergoes discontinuities in some surface in the system state space [44].

References

1. M.R. Anbiyaei, *White Noise Reduction for Wideband Sensor Array Signal Processing* (University of Sheffield, 2018)
2. S. Manikandan, Literature survey of active noise control systems. Acad. Open Internet J. **17** (2006)
3. A. Azizi, Computer-based analysis of the stochastic stability of mechanical structures driven by white and colored noise. Sustainability **10**(10), 3419 (2018)
4. A. Ashkzari, A. Azizi, Introducing genetic algorithm as an intelligent optimization technique, in *Applied Mechanics and Materials*, vol. 568. (Trans Tech Publ, 2014), pp. 793–797
5. A. Azizi, Introducing a novel hybrid artificial intelligence algorithm to optimize network of industrial applications in modern manufacturing. Complexity **2017** (2017)

6. A. Azizi, Hybrid artificial intelligence optimization technique, in *Applications of Artificial Intelligence Techniques in Industry 4.0* (Springer, 2019), pp. 27–47
7. A. Azizi, Modern manufacturing, in *Applications of Artificial Intelligence Techniques in Industry 4.0* (Springer, 2019), pp. 7–17
8. A. Azizi, RFID network planning, in *Applications of Artificial Intelligence Techniques in Industry 4.0* (Springer, 2019), pp. 19–25
9. A. Azizi, *Applications of Artificial Intelligence Techniques in Industry 4.0* (ed: Springer)
10. A. Azizi, F. Entesari, K.G. Osgouie, M. Cheragh, Intelligent mobile robot navigation in an uncertain dynamic environment, in *Applied Mechanics and Materials*, vol. 367. (Trans Tech Publ, 2013), pp. 388–392
11. A. Azizi, F. Entessari, K.G. Osgouie, A.R. Rashnoodi, Introducing neural networks as a computational intelligent technique, in *Applied Mechanics and Materials*, vol. 464. (Trans Tech Publ, 2014), pp. 369–374
12. A. Azizi, N. Seifipour, *Modeling of Dermal Wound Healing-Remodeling Phase by Neural Networks*, in International Association of Computer Science and Information Technology-Spring Conference, 2009, IACSITSC'09 (IEEE, 2009) pp. 447–450
13. A. Azizi, A. Vatankhah Barenji, M. Hashmipour, Optimizing radio frequency identification network planning through ring probabilistic logic neurons. Adv. Mech. Eng. **8**(8), p. 1687814016663476 (2016)
14. A. Azizi, P.G. Yazdi, M. Hashemipour, Interactive design of storage unit utilizing virtual reality and ergonomic framework for production optimization in manufacturing industry. Int. J. Interac. Des. Manuf. (IJIDeM) 1–9 (2018)
15. M. Koopialipoor, A. Fallah, D.J. Armaghani, A. Azizi, E.T. Mohamad, Three hybrid intelligent models in estimating flyrock distance resulting from blasting. Eng. Comput. 1–14 (2018)
16. K.G. Osgouie, A. Azizi, *Optimizing Fuzzy Logic Controller for Diabetes Type I by Genetic Algorithm*. in The 2nd International Conference on Computer and Automation Engineering (ICCAE), 2010, vol. 2. (IEEE, 2010), pp. 4–8
17. S. Rashidnejhad, A.H. Asfia, K.G. Osgouie, A. Meghdari, A. Azizi, Optimal trajectory planning for parallel robots considering time-jerk, in *Applied Mechanics and Materials*, vol. 390. (Trans Tech Publ, 2013), pp. 471–477
18. S. Elliott, *Signal Processing for Active Control* (Elsevier, 2000)
19. P.A. Nelson, S.J. Elliott, Active noise control: a tutorial review. IEICE Trans. Fundam. Electron. Commun. Comput. Sci. **75**(11), 1541–1554 (1992)
20. S.V. Vaseghi, *Advanced Digital Signal Processing and Noise Reduction* (Wiley, 2008)
21. S.C. Douglas, Introduction to adaptive filters, in *Digital Signal Processing Handbook* (1999), pp. 7–12
22. S. Das, K.K. Sarma, Noise cancellation in stochastic wireless channels using coding and adaptive filtering. Communicated Int. J. Comput. Appl. (IJCA), 2012
23. S.O. Haykin, *Adaptive Filter Theory* (Pearson Higher Ed, 2013)
24. F.R. Jiménez-López, C.E. Pardo-Beainy, E.A. Gutiérrez-Cáceres, Adaptive filtering implemented over TMS320c6713 DSP platform for system identification. Iteckne **11**(2), 157–171 (2014)
25. R.K. Ravi, *FPGA Implementation of Adaptive Filter Architectures* (2012)
26. M. Liu, C. Suh, Simultaneous time-frequency control of friction-induced instability. J. Appl. Nonlinear Dyn. **3**(3), 227–243 (2014)
27. A.P. Vinod, E.M.-K. Lai, in *Design of Low Complexity High-Speed Pulse-Shaping IIR Filters for Mobile Communication Receivers*. IEEE International Symposium on Circuits and Systems, 2005, ISCAS 2005 (IEEE, 2005), pp. 352–355
28. B. Friedlander, Lattice filters for adaptive processing. Proc. IEEE **70**(8), 829–867 (1982)
29. V.J. Mathews, Adaptive polynomial filters. IEEE Signal Process. Mag. **8**(3), 10–26 (1991)
30. S.M. Kuo, D.R. Morgan, in *Review of DSP Algorithms for Active Noise Control*. Proceedings of the 2000 IEEE International Conference on Control Applications (IEEE, 2000), pp. 243–248
31. S.M. Kuo, D.R. Morgan, Active noise control: a tutorial review. Proc. IEEE **87**(6), 943–973 (1999)

32. B. Widrow et al., Adaptive noise cancelling: principles and applications. Proc. IEEE **63**(12), 1692–1716 (1975)
33. Z. Ren, Y. Zou, Z. Zhang, Y. Hu, in *Fast Extraction of Somatosensory Evoked Potential Using RLS Adaptive Filter Algorithms*. 2nd International Congress on Image and Signal Processing, 2009, CISP'09 (IEEE, 2009), pp. 1–4
34. K. Talele, A. Shrivastav, K. Utekar, A. Deshpande, in *LMS filter for Noise Cancellation Using Simulink*. Third International Conference on Digital Image Processing (ICDIP 2011), vol. 8009. (International Society for Optics and Photonics, 2011), p. 80093 K
35. M.M. Mahajan, S. Godbole, Design of least mean square algorithm for adaptive noise canceller. Int. J. Adv. Eng. Sci. Technol. **5**, 172–176 (2011)
36. H.-M. Park, S.-H. Oh, S.-Y. Lee, Adaptive noise cancelling based on independent component analysis. Electron. Lett. **38**(15), 832–833 (2002)
37. A. Sahu, S. K. Hota, in *Performance Comparison of 2-DOF PID Controller Based on Moth-flame Optimization Technique for Load Frequency Control of Diverse Energy Source Interconnected Power System*. Technologies for Smart-City Energy Security and Power (ICSESP) (IEEE, 2018), pp. 1–6
38. H. Senberber, A. Bagis, in *Fractional PID Controller Design for Fractional Order Systems Using ABC Algorithm*. Electronics, 2017 (IEEE, 2017), pp. 1–7
39. K. Jagatheesan, B. Anand, K.N. Dey, A.S. Ashour, S.C. Satapathy, Performance evaluation of objective functions in automatic generation control of thermal power system using ant colony optimization technique-designed proportional–integral–derivative controller. Electr. Eng. **100**(2), 895–911 (2018)
40. B. Yaghooti, H. Salarieh, Robust adaptive fractional order proportional integral derivative controller design for uncertain fractional order nonlinear systems using sliding mode control. Proc. Inst. Mech. Eng., Part I: J. Syst. Control Eng. **232**(5), 550–557 (2018)
41. W. Liao, Z. Liu, S. Wen, S. Bi, D. Wang, in *Fractional PID Based Stability Control for a Single Link Rotary Inverted Pendulum*. International Conference on Advanced Mechatronic Systems (ICAMechS), 2015 (IEEE, 2015), pp. 562–566
42. R. Munje, B. Patre, A. Tiwari, Sliding mode control, in *Investigation of Spatial Control Strategies with Application to Advanced Heavy Water Reacto*. (Springer Singapore, Singapore, 2018), pp. 79–91
43. R.A. DeCarlo, S.H. Zak, G.P. Matthews, Variable structure control of nonlinear multivariable systems: a tutorial. Proc. IEEE **76**(3), 212–232 (1988)
44. J. Guldner, V.I. Utkin, Sliding mode control for gradient tracking and robot navigation using artificial potential fields. IEEE Trans. Robot. Autom. **11**(2), 247–254 (1995)

Chapter 6
Modeling and Control of the Effect of the Noise on the Mechanical Structures

6.1 Overview

The goal of this research is to design two effective proportional–integral–derivative (PID) and sliding mode controllers, which will control the active suspension system of a car, in order to eliminate the imposed vibration to the car from pavement. In this research, Gaussian white noise has been adopted to model the pavement condition, and MATLAB software which has been adopted to perform simulation in many researches [1–15] has been utilized to model the effect of the Gaussian white noise on quarter car model, as well as to design effective controllers to generate the active force in active suspension system. The results show that the designed controllers are effective in eliminating the effect of road conditions. This has a significant effect on reducing the fuel consumption and contributes to environmental sustainability.

In this chapter, the quarter car dynamic vibration model has been chosen to represent the active suspension system (Fig. 6.1). While this model has limitations, such as eliminating vehicle's pitching and roll angle vibrations, it also includes the most essential features for this research, like changing the information related to the stress and load of the suspension system. The proposed model has been utilized by many researchers to investigate the effect of pavement conditions on body vibration of a vehicle[16–20].

6.2 Active Suspension System

As shown in Fig. 6.1, the vehicle body mass known as the sprung mass has been shown with M_1, and the mass of the axel and wheel which has been shown with M_2 represents the unsprung mass. It has been considered that the vehicle tires surly contacting the road surface when it is moving on the road and it is considered as a linear spring with stiffness K_2 in the modeling. In order to modeling the active

Fig. 6.1 Proposed quarter
car model equipped
with Active suspension
system [1]

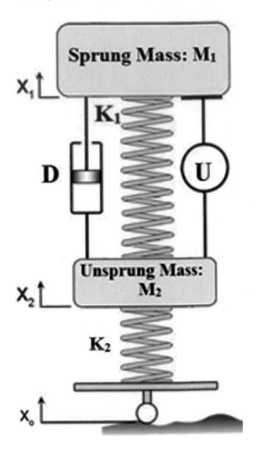

suspension system of the vehicle, a linear damper with D as the average damping
coefficient and a linear spring with K_1 as the average stiffness coefficient have been
considered. The vertical displacements of the M_1 and M_2, respectively, have been
represented by the state variables $X_1(t)$ and $X_2(t)$, since vertical pavement condition
has been shown by $X_0(t)$. The active control force which has been created by the
active suspension actuator is shown by U.

Generating a mathematical model is the first step of the procedure of modeling
a system, and it has been followed by calculating the design parameters. In control
engineering field, a system can be modeled mathematically in three different ways:
transfer function, weight function, and state space description [21].

In this book, the active quarter car suspension model which has been presented
in Fig. 6.1 has been modeled mathematically by transfer function method. To fulfill
the task, two degrees of freedom motion differential equations have been generated
as followed by analyzing the vehicle suspension system dynamics (Fig. 6.1) [1].

$$M_1\ddot{x}_1(t) + D[\dot{x}_1(t) - \dot{x}_2(t)] + k_1[x_1(t) - x_2(t)] = u \qquad (6.1)$$

$$M_2\ddot{x}_2(t) - D[\dot{x}_1(t) - \dot{x}_2(t)] + k_1[x_2(t) - x_1(t)] + k_2[x_2(t) - x_0(t)] = -u \quad (6.2)$$

One must assume that all of the initial conditions are zero, so these equations represent a situation when the wheel of a car goes over a bump. The dynamics of (6.1) and (6.2) assume that all initial conditions are zero, and there can be expressed in the form of transfer functions by taking Laplace transform of the equations. This is to represent the condition of when vehicle goes over a bump. It is important to know that the system will have two transfer functions, as represented in (6.3) and (6.4) [1]:

$$G_1(s) = \frac{x_1(s) - x_2(s)}{U(s)} = \frac{(M_1 + M_2)s^2 + k_2}{\Delta} \quad (6.3)$$

$$G_2(s) = \frac{x_1(s) - x_2(s)}{x_0(s)} = \frac{-M_1 k_2 + s^2}{\Delta} \quad (6.4)$$

where

$$\Delta = \det \begin{bmatrix} (m_1 s^2 + Ds + k_1) & -(Ds + k_1) \\ -(Ds + k_1) & (m_2 s_2 + Ds + (k_1 + k_2)) \end{bmatrix} \quad (6.5)$$

In fact, $G_1(s)$ represents the effect of exerted force on the vertical displacement of the car which has been produced by active suspension system, and $G_2(s)$ represents the effects of the pavement condition on the vertical displacement of the car. This means that vertical displacement of the vehicle is superposition of the effects of both active suspension force and pavement condition.

As mentioned in previous section, the goal of this research is to eliminate the effect of pavement condition on the system by utilizing a controller (in other words, by adjusting the produced force by active suspension, the effect of the road condition can be eliminated). To achieve this goal, a PID controller is proposed and explored in next section, in order to control the amount of the produced force.

Now by knowing the active suspension model and PID controller concept, the next step is to design an effective PID controller for the quarter car model.

6.3 Design of PID Controller

The goal of this chapter is to design an effective controller by eliminating the effect of the pavement on the vehicle passengers. The controller should be designed to make its system stable, by eliminating the disturbance of the road, which shows itself as an oscillation of the vehicle; and, at the same time, has a smooth and fast control signal to ensure that passengers' comfort and safety are not compromised. PID controllers are one of the best controllers that can be utilized for this purpose,

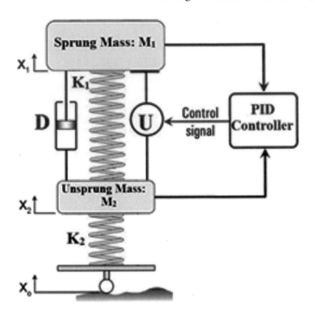

Fig. 6.2 Active suspension system equipped with PID controller [1]

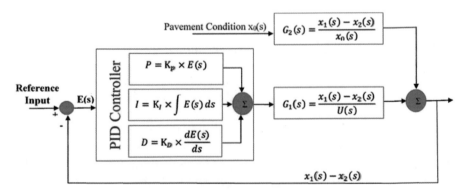

Fig. 6.3 Block diagram of the proposed model equipped with PID controller [1]

since they are capable to reach the steady state error by having a short rise time, and they can give stability to a system by eliminating oscillations and overshoot of the system (Fig. 6.2).

The proposed PID controller in this book consists of all three controller modes: proportional, integral, and derivative. The proposed controller is capable of eliminating the noise of the pavement by adjusting the force of proposed active suspension system (see Fig. 6.3).

It is important to know that how the proposed PID controller is working:

As mentioned previously, in this research the effect of the exerted force on the vertical displacement of the car produced by active suspension system has been mathematically modeled as $G_1(s)$. The input of this mathematical model is the PID controller output which results in the displacement of body of the vehicle (sprung mass) with respect to the unsprung part. However, there is one more element which affects this displacement, such as the pavement condition. The road condition shows its effect as disturbance on the system, and in this chapter, it has been modeled mathematically as $G_2(s)$.

Since the reference input should be equal to zero, it can be concluded that the difference between the reference input and total displacement which is known as error function $E(s)$ is the superposition of the effects of the control signal and noise signal on vehicle. In this case, the PID controller adjusts the active suspension force by calculating each mode signal based on the error function $E(s)$ (Fig. 6.3).

The proposed PID controller for the modeled quarter car's active suspension system has been simulated in MATLAB–Simulink environment. The simulation has seen performed based on the parameters which can be seen in Table 4.2.

6.3.1 Results

As seen in Fig. 6.4, the proposed methodology has been implemented and modeled in MATLAB–Simulink environment. The model can be dived in two main parts: controlled and uncontrolled parts. The uncontrolled part consists of the effect of the pavement conditions on vehicle's vertical displacement and the effect of these vibrations on the fuel consumption rate. The controlled part consists of the proposed PID controller to reduce the effect of the imposed fluctuations by the pavement condition on the vehicle and the effect of the controller on reduction of the fuel consumption rate. The pavement condition $x_0(s)$ has been modeled by Gaussian white noise. The reason is that the road unevenness is kind of noise which should be compensated, so any kind of colored noise such as pink, red, and green can be utilized for this purpose [22]. Since white noise contains all frequencies of colored noises, it is a good approximation to simulate the randomness of pavement roughness [23].

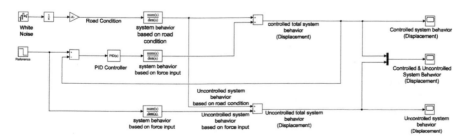

Fig. 6.4 MATLAB–Simulink model of the noise cancelation system for the active suspension system

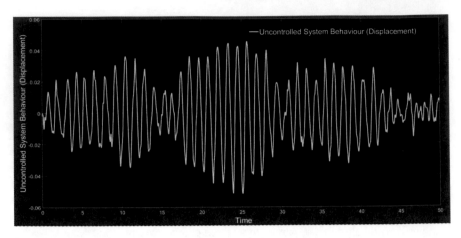

Fig. 6.5 Uncontrolled total system response based on the random input [1]

As seen in Fig. 6.4, the reference input which here is the desired output has been simulated by step function. The reason is that the reducing and even eliminating the vehicle oscillations is desirable, and the oscillations are resulted by displacements of sprung mass with respect to unsprung mass. The displacement has been calculated as plant's output $(x_1(s)-x_2(s))$, since the target is to eliminate it, so the desired output should be considered as zero, and a step function which generating zero value at $t > 0$ is a good model to generate the required data.

The effect of the random road pavement which has been modeled by Gaussian white noise generator has been illustrated in Fig. 6.5. It can be observed that the vehicle follows the road condition and fluctuates with road fluctuations which can be considered as unstable behavior and will result in increasing fuel consumption, decreasing effective life of the vehicle's part, and compromising passengers' safety.

As seen in Fig. 6.6, this uncontrolled response is not convenient for the passengers, so to reduce and eliminate the effect of the road pavement on the car which shows itself as car fluctuation, as described before an PID controller has been designed.

By utilizing the PID controller as it can be observed from Fig. 6.6, the car does not follow the road condition and active suspension controller cancels out the unpleasant pavement condition's effect on the car.

To have a better understanding that how the proposed PID controller compensates the effect of the pavement, the total response of the system before and after utilizing the PID controller has been shown in Fig. 6.7. As it can be observed system shows an aggressive controlled response to the input noise in first two seconds of the simulation steps, but after three seconds it starts to cancel the noise and prepares a convenient and safe ride for passengers.

To explore the stability status of the designed controller in this research, as it can be seen in Fig. 6.8, the compensator editor tool of the MATLAB software has been adopted. The system, without considering the forced step function input, has only one negative pole at -100 and two complex zeroes at $-0.405 \pm 1.19i$. If the

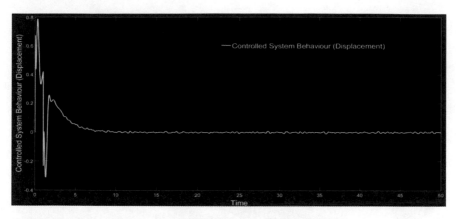

Fig. 6.6 Controlled total system response based on the random input [1]

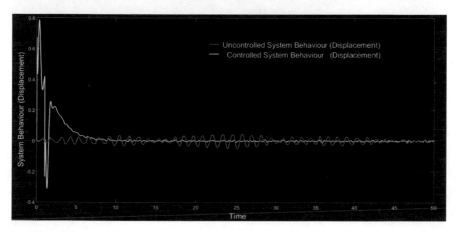

Fig. 6.7 Controlled and uncontrolled total system behaviors [1]

forced input effect is desired for consideration, another pole in zero must be added to the system. Since the system only has a one negative pole, it can be concluded that the designed PID controller (based on the location of the poles) is a stable system, and (based on the location of the zeros) it has a damping ratio of 0.3, and a natural frequency of 1.25 Hz.

Now by knowing the locations of the pole and zeros of the system, root locus diagram can be plot. As seen in Fig. 6.9, it can be concluded that by moving to the left side of the real axis, the system shows more stable behavior.

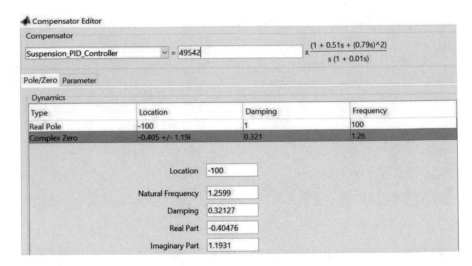

Fig. 6.8 Pole and zero of the designed PID controller [1]

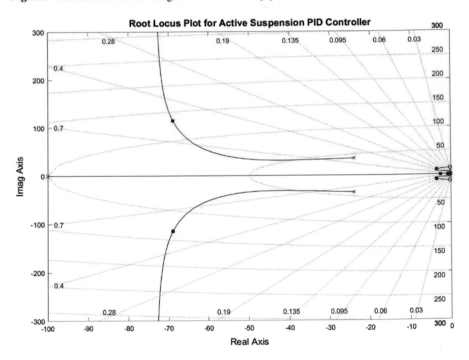

Fig. 6.9 Root locus diagram [1]

6.4 Design of Sliding Mode Controller

This section aims to design an effective sliding mode controller for the active suspension system discussed in previous section.

Sliding mode control is one of the most efficient techniques for controlling of nonlinear systems [24, 25]. It is also highly utilizing for robust control of nonlinear and linear systems. Sliding mode control is functioning based on designing a steady sliding surface. The reason behind the need for this design is to settle the dynamical and states error on this surface. In the other word, the sliding surface is acting as an attraction set which can absorb the error or the controlled states. Due to the uncertainty and noise presence, the mentioned error or the controlled conditions can never reach to zero, but they will maintain in steady range.

Equation, related to the vibration of the suspension system which is illustrated in Fig. 6.1, is stated in (6.6). It is possible to rewrite the mentioned equation into (6.7).

$$\begin{bmatrix} M_1 & 0 \\ 0 & M_2 \end{bmatrix}\begin{bmatrix} \ddot{x}_1 \\ \ddot{x}_2 \end{bmatrix} + \begin{bmatrix} k_1 & -k_1 \\ -k_1 & k_1 + k_2 \end{bmatrix}\begin{bmatrix} x_1 \\ x_2 \end{bmatrix} + \begin{bmatrix} D & -D \\ -D & D \end{bmatrix}\begin{bmatrix} \dot{x}_1 \\ \dot{x}_2 \end{bmatrix} = \begin{bmatrix} u \\ -u \end{bmatrix} + \begin{bmatrix} 0 \\ k_2 x_0 \end{bmatrix} \tag{6.6}$$

$$M_1\ddot{x}_1 + k_1 x_1 - k_1 x_2 + D\dot{x}_1 - D\dot{x}_2 = u$$
$$M_2\ddot{x}_2 - k_1 x_1 + (k_1 + k_2)x_2 - D\dot{x}_1 + D\dot{x}_2 = -u + k_2 x_0 \tag{6.7}$$

The main control objective is to stabilize $x_1 - x_2$. So that a new state is defined and is named as e which is representing the error. As the result of this definition, the dynamic error is

$$\dot{e} = \dot{x}_1 - \dot{x}_2 \tag{6.8}$$

For this step is tried to represent the error in the state space. $e = e_1$ and $\dot{e} = e$ is assumed and generally, we can get the following result:

$$\begin{cases} e = x_1 - x_2 \\ \dot{e} = \dot{x}_1 - \dot{x}_2 \end{cases} \rightarrow \begin{cases} e = e_1 \\ \dot{e}_1 = e_2 \end{cases} \tag{6.9}$$

And, also:

$$\begin{cases} \dot{e}_1 = e_2 \\ \dot{e}_2 = \ddot{e}_1 = \ddot{x}_1 - \ddot{x}_2 \end{cases} \tag{6.10}$$

The $\ddot{x}_1$ and $\ddot{x}_2$ can be extracted from the dynamics of the car suspension system. So that

$$\begin{cases} \ddot{x}_1 = \frac{1}{M_1}(u - k_1 e_1 - De_2) \\ \ddot{x}_2 = \frac{1}{M_2}(k_1 e_1 + De_2 - k_2 x_2 - u + k_2 x_0) \end{cases} \tag{6.11}$$

The general form of the dynamics of error in the state space will be obtained by merging the (6.10) and (6.11)

$$\begin{cases} \dot{e}_1 = e_2 \\ \dot{e}_2 = \frac{1}{M_1}(u - k_1 e_1 - De_2) - \frac{1}{M_2}(k_1 e_1 + De_2 - k_2 x_2 - u + k_2 x_0) \end{cases} \tag{6.12}$$

The sliding surface for designing the sliding mode controller will be considered as $s = e_2 + \lambda e_1$. Here, λ represent a positive number. So that, the sliding surface and it dynamic is

$$\begin{cases} s = e_2 + \lambda e_1 \\ \dot{s} = \dot{e}_2 + \lambda \dot{e}_1 = \dot{e}_2 + \lambda e_2 \end{cases} \tag{6.13}$$

By substitution of the (6.12) into (6.13), the final form of the dynamic of the sliding surface will be achieved as:

$$\dot{s} = \frac{1}{M_1}(u - k_1 e_1 - De_2) - \frac{1}{M_2}(k_1 e_1 + De_2 - k_2 x_2 - u + k_2 x_0) + \lambda e_2 \tag{6.14}$$

6.4.1   Stability in the Sense of Lyapunov

With considering the sense of Lyapunov, the $x^* = 0$ which is the equilibrium point is stable at $t = t_0$ if there is $\delta(t_0, \varepsilon) > 0$ for any $\varepsilon > 0$ (see Eq. 6.15) [26].

$$\|x(t_0)\| < \delta \Rightarrow \|x(t)\| < \varepsilon, \forall t \geq t_0 \tag{6.15}$$

Lyapunov stability analysis is a mild requirment on equilibrium points. It means that, it is not neccessary that trajectories to start close to the origin tend to the origin asymptotically. In addition, at t_0 time instant we can define the stability. Stability of the equilibrium point is guaranteed by the concept of uniform stability. It is important to mention that for a uniformly stable equilibrium point x^*, δ is not a function of t_0, so that (6.15) may hold for all t_0 [26]. In order to generalize the (6.15) for all of the t_0 and having the uniform stability of the equilibrium point (x^*), δ should not be a t_0 function [27].

6.4.2   Asymptotic Model

Whenever equilibrium point $x^* = 0$ is locally attractive and is stable, then it is asymptotically stable at $t = t_0$; For example, if $\delta(t_0)$ exists as follows [26, 28]:

$$\|x(t_0)\| < \delta \Rightarrow \lim_{t \to \infty} x(t) = 0 \tag{6.16}$$

In addition to previous state, if $x^* = 0$ is locally attractive and also uniformly stable, then it is uniformly asymptotic stabile. For example, in (6.16), if δ exists independently of t_0 and there is the uniform convergence. Generally, in contrast to the above arguments, if it is not stable, the equilibrium point will be unstable.

Stability in both cases of asymptotic and in the sense of Lyapunov can be defined locally; the system's behavior will be described close to the equilibrium point. If the equilibrium point x^* is stable for all $x_0 \in R_n$ as initial conditions, then it is stable globally. In most of the cases and applications, global stability is hard to achieve and is difficult. Here the local stability is focused, and the results can be generalized globally. Uniformity will be important when we are facing with time-varying system. Therefore, for this type of systems, stability indicates as uniform stability and asymptotic stability indicates as uniform asymptotic stability. The term of asymptotic stability is not covering the convergence rate.

6.4.3 Lyapunov's Direct/Second Method

In the direct mode or second method of Lyapunov determination of the stability of the system is not depending on (6.15). This method is based on the measurement of the energy in the system and studying the change rate of energy to investigate the stability [29]. In order to achieve this, "measure of energy" should be defined properly.

With assuming the B_ε around the origin as a ball, then we have the following [29]:

$$B_\varepsilon = \left\{ x \in R^n : \|x\| < \varepsilon \right\}. \tag{6.17}$$

In this section, some functions have been defined to clarify the Lyapunov direct mode.

6.4.3.1 Locally Positive Definite Functions

Locally positive definite function can be defined as following [27]:

A continuous function $V : R^n \times R_+ \to R$ is a locally positive definite function if for some $\varepsilon > 0$ and some continuous, strictly increasing function $\alpha : R_+ \to R$,

$$V(0, t) = 0 \text{ and } V(x, t) \geq \alpha(\|x\|) \tag{6.18}$$

$$\forall x \in B_\varepsilon$$
$$\forall t \geq 0$$

The energy function is similar to a locally positive definite function. Positive definite functions are actually the ones like energy functions [27].

6.4.3.2 Positive Definite Functions

A continuous function $V : R^n \times R_+ \rightarrow R$ is a positive definite function if it satisfies the conditions of locally positive definite function and, additionally, $\alpha(p) \rightarrow \infty$ as $p \rightarrow \infty$. To bound the energy function from above, decrescent should be defined [27].

6.4.3.3 Decrescent Functions

A continuous function $V : R^n \times R_+ \rightarrow R$ is decrescent if for some $\varepsilon > 0$ and some continuous, strictly increasing function $\beta : R_+ \rightarrow R$ (see Eq. 6.19) [27].

$$V(x, t) \leq \beta(\|x\|)$$
$$\forall x \in B_\varepsilon,$$
$$\forall t \geq 0 \tag{6.19}$$

By utilizing the following equation, we will be able to evaluate the stability of the system through a proper energy function. Roughly, this theorem states that the equilibrium point is stable when $\dot{V}(x, t) \leq 0$ and $V(x, t)$ is a locally positive definite function. The time derivative of V can be calculated as follows [30]:

$$\dot{V}\Big|_{\dot{x}=f(x,t)} = \frac{\partial V}{\partial t} + \frac{\partial V}{\partial x} f \tag{6.20}$$

Summary of the basic theorem of Lyapunov is shown in Table 6.1.

With utilizing the direct mode of Lyapunov, the dynamics system's stability can be evaluated easily because in this method system equations are not going to be solved necessarily. The following section briefly explains this method for the deterministic systems [31].

Consider $V(x)$ which is defined on domain R^n, and it is a positive and continuous function. Suppose for some $m \in R$, the set $Q_m = \{x \in R^n : V(x) < m\}$ is bounded and $V(x)$ has continuous first partial derivatives in Q_m. Let the initial time $t_0 = 0$ and let $x(t) = x(t, x0)$ be the unique solution of the initial value problem as follows [31]:

$$\begin{cases} \dot{x}(t) = f(x(t)), t \geq 0, \\ x(0) = x_0 \in R^n, f(0) = 0, \end{cases} \tag{6.21}$$

for $x_0 \in Q_m$. Because $V(x)$ is continuous, the open set Q_r for r $\in [0, m]$ defined by $Q_r = \{x \in R^n : V(x) < r\}$ contains the origin and monotonically decreases to the

Table 6.1 Basic theorem of Lyapunov [26]

	Conditions on $V(x, t)$	Conditions on $-\dot{V}(x, t)$	Conclusions
1	Locally positive definite function	≥ 0 locally	Stable
2	Locally positive definite function, decrescent	≥ 0 locally	Uniformly stable
3	Locally positive definite function, decrescent	Locally positive definite function	Uniformly asymptotically stable
4	Locally positive definite function, decrescent	Locally positive definite function	Globally uniformly asymptotically stable

singleton set $\{0\}$ as $r \rightarrow 0^+$. The total derivative $\dot{V}(x)$ of $V(x)$ is given as follows [31]:

$$\dot{V} = \frac{dV(x)}{dt} = f^T(x) \cdot \frac{\partial V}{\partial x} \overset{\text{def}}{=} -k(x) \qquad (6.22)$$

If $k(x)$ be considered as positive and continuous function, then it can be observed that for all the time greater than zero always $V(x(t))$ will be less than m. It is important to know that the Lyapunov stability of the zero solution of (6.21) can be obtained by function $V(x)$ which is the definition of the Lyapunov function. Since $k(x)$ is a positive and continuous function, it can be observed from (6.22) that $V(x(t)) \rightarrow 0$ as $t \rightarrow +\infty$. It means that $V(x(t))$ is a monotone decreasing function, also integration of (6.22) shows the same result, as follows [31]:

$$0 < V(x_0) - V(x(t)) = \int_0^t k(x_s)ds < +\infty \quad \text{for } t \in [0, +\infty) \qquad (6.23)$$

The mentioned state shows that if the physical system's energy is decreasing continually near an equilibrium point, then the stability of equilibrium state is approved.

As the disadvantage of Lyapunov function, we can state that generally for nonlinear systems categories, there is no systematic method available for creating a proper Lyapunov function, and it is affecting the stability criteria determination because it critically depends on the Lyapunov function.

In order to continue the procedure, the sliding mode controller will be discussed. A Lyapunov function in the form of $v = \frac{1}{2}s^2$ is considered. To guarantee the stability of the sliding mode controller, the Lyapunov time derivative should be a negative value ($\dot{v} < 0$). With operating the time derivative of the $v = \frac{1}{2}s^2$, we have

$$\dot{v} = s\dot{s} \tag{6.24}$$

Finally, the general definition of the sliding mode controller design will be

$$v = \frac{1}{2}s^2$$
$$\dot{v} = s\dot{s} < 0 \tag{6.25}$$

In this phase by substitution of $\dot{s}$ from (6.14) into (6.25), the related equation to $\dot{v}$ can be obtained as

$$\dot{v} = s\left[\frac{1}{M_1}(u - k_1e_1 - De_2) - \frac{1}{M_2}(k_1e_1 + De_2 - k_2x_2 - u + k_2x_0) + \lambda e_2\right] \tag{6.26}$$

The simple form of the $\dot{v}$ equation will be presented as

$$\dot{v} = s\left[\begin{array}{c} -k_1e_1\left(\dfrac{1}{M_1} + \dfrac{1}{M_2}\right) - De_2\left(\dfrac{1}{M_1} + \dfrac{1}{M_2} + \dfrac{\lambda}{D}\right) + \dfrac{k_2}{M_2}(x_2 - x_0) \\[2mm] +u\left(\dfrac{1}{M_1} + \dfrac{1}{M_2}\right) \end{array}\right] \tag{6.27}$$

Equation (6.28) is representation of a true relation, and it is the foundation of control parameter definition.

$$\dot{v} \le |s|\left[k_1Q|e_1| + D\left(Q + \frac{\lambda}{D}\right)|e_2| + \frac{k_2}{M_2}|x_2 - x_0|\right] + suQ = -\eta|s| \tag{6.28}$$

where $Q = \frac{1}{M_1} + \frac{1}{M_2}$ and η is a positive number.
f is appositive defined function and will be represented as:

$$f = k_1Q|e_1| + D\left(Q + \frac{\lambda}{D}\right)|e_2| + \frac{k_2}{M_2}|x_2 - x_0| \tag{6.29}$$

With considering the f and its substitution in the right side of (6.28), we can obtain

$$suQ = (-\eta - f)|s| \tag{6.30}$$

and the control parameter will be obtained as

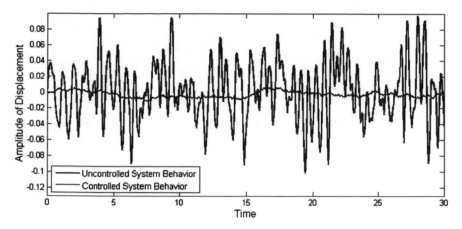

Fig. 6.10 Controlled and uncontrolled system behaviors for $\eta = 1$ and $\lambda = 1$

$$u = \frac{(-\eta - f)}{Q} \frac{|s|}{s}$$

or

$$u = \frac{(-\eta - f)}{Q} \mathrm{sign}(s) \tag{6.31}$$

And sign (s) is defined as following equation:

$$\mathrm{sign}(s) = \mathrm{if} \begin{cases} s > 0 \rightarrow s = 1 \\ s = 0 \rightarrow s = 0 \\ s < 0 \rightarrow s = -1 \end{cases} \tag{6.32}$$

In the rest of this section, the result of the control of the mentioned car suspension system is presented. The specific characteristics of the system are presented in Table 4.2.

The result of the controlled and uncontrolled system responses are illustrated in the 6.10. White noise has been considered as the uncontrolled system input. Also parameteres of the sliding mode are assumed as $\eta = 1$ and $\lambda = 1$.

As it is obvious in the Fig. 6.10, utilizing the sliding mode controller decreases the amplitude of unwanted system fluctuation.

In the following section, the effect of the variation of η and λ on the behavior of the system will be investigated. The system behavior with $\eta = 1$ and different λ is shown in Figs. 6.11 and 6.12.

Figures 6.10 and 6.11 show that the maximum amplitude of vibration decreases for $\lambda = 100$ and $\lambda = 1000$. The minimum amplitude of vibration in these figures belongs to $\lambda = 1000$.

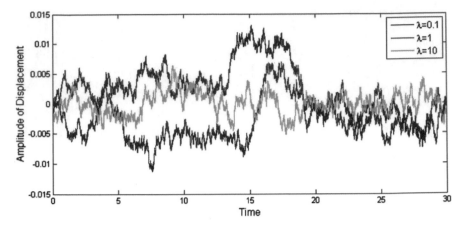

Fig. 6.11   Controlled system behavior for $\lambda = 0.1$, $\lambda = 1$, $\lambda = 10$

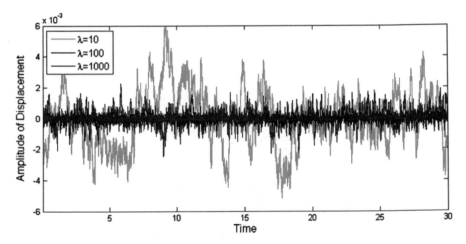

Fig. 6.12   Controlled system behavior for $\lambda = 10$, $\lambda = 100$, $\lambda = 1000$

In addition, the system behavior with $\lambda = 1$ and different η is shown in Figs. 6.13, 6.14, and 6.15.

As can be seen in Figs. 6.13, 6.14, and 6.15, the maximum amplitude of vibration is not visible for the $0.01 \leq \eta \leq 10000$. Although the amount of this increment for η between $0.01 < \eta < 10000$ is not that much apparent in the value of $\eta = 100000$, it is obviously increased. It is with considering that the vibration amplitude is even more than the uncontrolled system amplitude. So that it is suggested less amount of η for this simulation and more amount of λ being utilized to decrease the vibration amplitude.

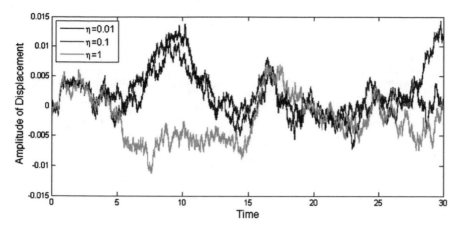

Fig. 6.13 Controlled system behavior for $\eta = 0.01,\ \eta = 0.1,\ \eta = 1$

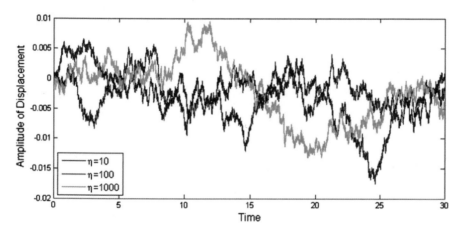

Fig. 6.14 Controlled system behavior for $\eta = 10,\ \eta = 100,\ \eta = 1000$

6.4.4 Conclusion

The main role of suspension systems is to reduce fuel consumption and to ensure passenger safety. The road roughness yields fluctuations of the vehicle wheels which will transmit to the all parts of the vehicle along as the passengers. It becomes clear that the role of the suspension system is to reduce as much vibrations and shocks occurring while driving time. An effective suspension system should result in a smooth driving with less vehicle vibrations and a degree of comfort based on the interaction with bumpy road surface. The vehicle behavior should not be with too large oscillations in the presence of a good suspension system. To achieve this goal in this chapter, an active car suspension has been modeled; two effective PID and sliding mode controllers to cancel the negative effects of the pavement conditions have been

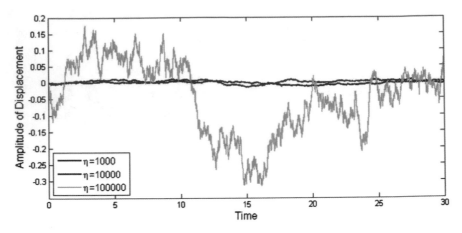

Fig. 6.15 Controlled system behavior for $\eta = 1000$, $\eta = 10,000$, $\eta = 100,000$

proposed and designed. Since the Gaussian white noise produces random outputs, it
has been adopted to simulate the pavement effects on the vehicle. Proposed plant and
control architecture has been modeled by MATLAB software package, and stability
of controllers has been investigated. The results show that the proposed controllers
work properly and have an effective performance which results in decreasing the
fuel consumption and preventing early vehicle's parts damage. As future work rather
than the proposed linear model, nonlinear elements can be considered for quarter car
model; also the PID controller's coefficients can be optimized by RPLNN, hybrid,
genetic algorithm, and other artificial intelligent technique.

References

1. A. Azizi, Computer-based analysis of the stochastic stability of mechanical structures driven by white and colored noise. Sustainability **10**(10), 3419 (2018)
2. A. Ashkzari, A. Azizi, Introducing genetic algorithm as an intelligent optimization technique. in *Applied Mechanics and Materials*, vol. 568 (Trans Tech Publication, 2014), pp. 793–797
3. A. Azizi, Introducing a novel hybrid artificial intelligence algorithm to optimize network of industrial applications in modern manufacturing. *Complexity* (2017)
4. A. Azizi, Hybrid artificial intelligence optimization technique. in *Applications of Artificial Intelligence Techniques in Industry 4.0*. (Springer, 2019), pp. 27–47
5. A. Azizi, Modern manufacturing. in *Applications of Artificial Intelligence Techniques in Industry 4.0*. (Springer, 2019), pp. 7–17
6. A. Azizi, RFID network planning. in *Applications of Artificial Intelligence Techniques in Industry 4.0*. (Springer, 2019), pp. 19–25
7. A. Azizi, Applications of artificial intelligence techniques in industry 4.0. (ed: Springer)
8. A. Azizi, F. Entesari, K.G. Osgouie, M. Cheragh, Intelligent mobile robot navigation in an uncertain dynamic environment. in *Applied Mechanics and Materials*, vol. 367. (Trans Tech Publication, 2013), pp. 388–392

9. A. Azizi, F. Entessari, K.G. Osgouie, A.R. Rashnoodi, Introducing neural networks as a computational intelligent technique. in *Applied Mechanics and Materials*, vol. 464, (Trans Tech Publication, 2014), pp. 369–374
10. A. Azizi N. Seifipour, Modeling of dermal wound healing-remodeling phase by Neural Networks. in *International Association of Computer Science and Information Technology-Spring Conference, 2009, IACSITSC'09.* (IEEE, 2009), pp. 447–450
11. A. Azizi, A. Vatankhah Barenji, M. Hashmipour, Optimizing radio frequency identification network planning through ring probabilistic logic neurons. Adv. Mech. Eng. **8**(8), 1687814016663476 (2016)
12. A. Azizi, P.G. Yazdi, M. Hashemipour, Interactive design of storage unit utilizing virtual reality and ergonomic framework for production optimization in manufacturing industry. Int J Interact Design Manuf (IJIDeM) 1–9 (2018)
13. M. Koopialipoor, A. Fallah, D.J. Armaghani, A. Azizi, E.T. Mohamad, Three hybrid intelligent models in estimating flyrock distance resulting from blasting. *Eng. Comput.* 1–14 (2018)
14. K.G. Osgouie A. Azizi, Optimizing fuzzy logic controller for diabetes type I by genetic algorithm. in *The 2nd International Conference on Computer and Automation Engineering (ICCAE), 2010* , vol. 2. (IEEE, 2010), pp. 4–8
15. S. Rashidnejhad, A.H. Asfia, K.G. Osgouie, A. Meghdari, A. Azizi, Optimal trajectory planning for parallel robots considering time-jerk, in *Applied Mechanics and Materials*, vol. 390. (Trans Tech Publication, 2013), pp. 471–477
16. M.S. Nagamani, S.S. Rao, S. Adinarayana, Minimization of human body responses due to automobile vibrations in quarter car and half car models using PID controller. SSRG Int J Mech Eng (SSRG-IJME) (2017)
17. M. Ma, P. Wang, C.-H. Chu, Redundant reader elimination in large-scale distributed RFID networks. IEEE Int Things J (2018)
18. P.M.K.P.P. Pravinkumar M.M.M. Mohanraj, Analysis of vehicle suspension system subjected to forced vibration using MAT LAB/Simulink (2017)
19. V. Barethiye, G. Pohit, A. Mitra, A combined nonlinear and hysteresis model of shock absorber for quarter car simulation on the basis of experimental data. Eng. Sci. Technol. Int. J. **20**(6), 1610–1622 (2017)
20. R. Guo, J. Gao, X.-K. Wei, Z.-M. Wu, S.-K. Zhang, Full vehicle dynamic modeling for engine shake with hydraulic engine mount. SAE Technical Paper 0148–7191 (2017)
21. D.J. Inman, *Vibration with control*. Wiley (2017)
22. P. Häunggi, P. Jung, Colored noise in dynamical systems. Adv. Chem. Phys. **89**, 239–326 (1994)
23. Y.O. Ouma, M. Hahn, Wavelet-morphology based detection of incipient linear cracks in asphalt pavements from RGB camera imagery and classification using circular Radon transform. Adv. Eng. Inform. **30**(3), 481–499 (2016)
24. J.-J.E. Slotine W. Li, *Applied nonlinear control* (no. 1) (Prentice hall Englewood Cliffs, NJ, 1991)
25. A. Isidori, *Nonlinear control systems* (Springer Science & Business Media, 2013)
26. R.M. Murray, *A mathematical introduction to robotic manipulation* (CRC press, 2017)
27. A.R. Teel, L. Praly, "A smooth Lyapunov function from a class-${\mathcal {KL}}$ estimate involving two positive semidefinite functions," ESAIM: Control. Optimisation Calculus Var **5**, 313–367 (2000)
28. N. Yeganefar, N. Yeganefar, M. Ghamgui, E. Moulay, Lyapunov Theory for 2-D Nonlinear Roesser Models: Application to Asymptotic and Exponential Stability. IEEE Trans. Automat. Contr. **58**(5), 1299–1304 (2013)
29. H. Leipholz, The Direct Method of Lyapunov. in *Stability Theory*. (Springer, 1987), pp. 77–88
30. J. Daafouz, J. Bernussou, Parameter dependent Lyapunov functions for discrete time systems with time varying parametric uncertainties. Syst. Control Lett. **43**(5), 355–359 (2001)
31. W.S. Levine, *The control systems handbook: Control system advanced methods* (CRC press, 2010)

Printed in the United States
By Bookmasters